METHODS IN MOLECULAR BIOLOGY™

Series Editor
John M. Walker
School of Life Sciences
University of Hertfordshire
Hatfield, Hertfordshire, AL10 9AB, UK

For other titles published in this series, go to
www.springer.com/series/7651

Nanoscale Biocatalysis

Methods and Protocols

Edited by

Ping Wang

Department of Bioproducts and Biosystems Engineering, Biotechnology Institute, University of Minnesota, St. Paul, MN, USA

Editor
Ping Wang
Department of Bioproducts
and Biosystems Engineering
Biotechnology Institute
University of Minnesota
St. Paul, MN 55108, USA
ping@umn.edu

ISSN 1064-3745 e-ISSN 1940-6029
ISBN 978-1-61779-131-4 e-ISBN 978-1-61779-132-1
DOI 10.1007/978-1-61779-132-1
Springer New York Dordrecht Heidelberg London

Library of Congress Control Number: 2011928153

Printed on acid-free paper

Humana Press is part of Springer Science+Business Media (www.springer.com)

Preface

In the pursuit of efficient and sustainable chemical processing technologies, people have seen a growing emphasis on synthetic biotechnology in recent R&D efforts. In particular, industrial enzyme technologies are attracting enormous attention. Having been traditionally developed for food processing and detergent applications, industrial enzyme technologies are being re-examined and tested to their capability limits in order to keep abreast of the challenges in drug, biochemical, and the emerging biorenewable energy industries. Toward that, enzymes are required to function in non-conventional conditions, such as organic solvents, extreme pH, and temperatures; they also have to compete against alternative chemical technologies in terms of costs and efficiency. Accordingly, enzymic biocatalyst systems are being tackled dynamically at all size levels through efforts ranging from molecular-level protein engineering and modification, nanoscale structure fabrication, intracellular reaction pathway controls and regulation, and microenvironment manipulation to the construction of micro-chip devices and macroscopic industrial bioreactors and devices.

Nanoscale science and engineering, which deal with size-dependent properties and phenomenon at nanometer scale, are unveiling new mechanisms that people have to rely on heavily nowadays to achieve such efficiencies. Advances in nanoscale science and technology are fueling a new wave of development in almost all the major areas of science and engineering such as electronics, sensors, energy harvesting and storage, and life science. Although it is hitherto only at a very early stage, their impetus on biocatalysis is evidently powerful and greatly promising.

To capitalize on the power of nanoscale technologies, it is becoming a common need to understand and design complex systems of multi-scale structures. In most cases, natively evolved biocatalysts, either in the form of enzymes or microbes, are not optimized for use in industrial reactors. Recent advances in genetic engineering have made it possible to design and produce more efficient industrial biocatalysts. The use of isolated enzymes provides the chance to further improve the performance of biocatalysts through various after-isolation chemical and physical manipulations. Nano-structured materials provide desirable features in balancing the key factors that determine the efficiency of biocatalysts, including specific surface area, mass transfer resistance, and effective enzyme loading. Various nanomaterials such as nanoparticles, nanofibers, nanotubes, and nanoporous matrices have demonstrated promising potentials in revolutionizing the preparation and use of biocatalysts. Extending from there, people are exploring broadly enzyme-based assemblies, micro-scale structures, and new types of bioreactors for biotransformation and biosynthesis. It is probably reasonable to say that synthetic biotechnology is entering a new phase of development that integrates biocatalysts' structural and functional features that exhibit at different size scale, ranging from atomic–molecular-level, nanometer structure scale to a degree of micro- and macro-size bioreactors.

This current book gathers methodologies and procedures that have been developed from recent research in this burgeoning area of nanoscale technology-enabled biocatalysis. Chapters 1 and 2 demonstrate concepts in preparing unique and dynamic protein structures for biocatalysis. Methods for preparation of enzyme assembles or complexes that maintain molecular-like Brownian mobility are provided in Chapters 3, 4, and 5. Chapters 6 through 15 show examples in developing protein–nanostructure complexes using carbon nanotubes (CNTs) and nanoparticles. The last three chapters describe methodologies that have great potential for scale-up preparation of nano-structured biocatalysts.

Ping Wang

Contents

Contributors

FRASER A. ARMSTRONG • *Inorganic Chemistry Laboratory, University of Oxford, Oxford, UK*

SHYAM SUNDHAR BALE • *Department of Chemical and Biological Engineering, Rensselaer Polytechnic Institute, Center for Biotechnology and Interdisciplinary Studies, Troy, NY, USA*

ANDREAS BUTHE • *Department of Bioproducts and Biosystems Engineering, BioTechnology Institute, University of Minnesota, St. Paul, MN, USA*

JONATHAN W. CHIN • *Department of Chemical Engineering, The Pennsylvania State University, University Park, PA, USA*

SAMJIN CHOI • *Department of Biomedical Engineering, College of Medicine, Kyung Hee University, Seoul, Republic of Korea; Healthcare Industry Research Institute, Kyung Hee University, Seoul, Republic of Korea*

SEOK KEUN CHOI • *Department of Neurosurgery, Kyung Hee University Medical Center, Seoul, Republic of Korea*

PATRICK C. CIRINO • *Department of Chemical Engineering, The Pennsylvania State University, University Park, PA, USA*

JUAN C. CRUZ • *Department of Chemical Engineering, Kansas State University, Manhattan, KS, USA*

PETER CZERMAK • *Department of Chemical Engineering, Kansas State University, Manhattan, KS, USA; University of Applied Sciences Giessen-Friedberg, Giessen, Germany*

CERASELA ZOICA DINU • *Department of Chemical and Biological Engineering, Rensselaer Polytechnic Institute, Center for Biotechnology and Interdisciplinary Studies, Troy, NY, USA; Department of Chemical Engineering, College of Engineering and Mineral Resources, West Virginia University, Morgantown, WV, USA*

JONATHAN S. DORDICK • *Department of Chemical and Biological Engineering, Rensselaer Polytechnic Institute, Center for Biotechnology and Interdisciplinary Studies, Troy, NY, USA*

MICHAEL C. FLICKINGER • *Departments of Microbiology; Chemical and Biomolecular Engineering, Golden-LEAF Biomanufacturing Training and Education Center, North Carolina State University, Raleigh, NC, USA*

FEI GAO • *National Key Laboratory of Biochemical Engineering, Institute of Process Engineering, Chinese Academy of Sciences, Beijing, China*

JUN GE • *Department of Chemical Engineering, Tsinghua University, Beijing, China*

JIMMY L. GOSSE • *Department of Biological and Agricultural Engineering, North Carolina State University, Raleigh, NC, USA*

MASAHIRO GOTO • *Department of Applied Chemistry, Graduate School of Engineering, Kyushu University, Fukuoka, Japan*

HIDEHIKO HIRAKAWA • *Department of Bioengineering, Graduate School of Engineering, Center for NanoBio Integration, The University of Tokyo, Tokyo, Japan*

HONGFEI JIA • *Materials Research Department, Toyota Research Institute of North America, Ann Arbor, MI, USA*

RONGRONG JIANG • *School of Chemical & Biomedical Engineering, Nanyang Technological University, Singapore, Singapore*
MARTIN KRAMER • *Department of Chemical Engineering, Kansas State University, Manhattan, KS, USA; Abbott Diagnostics Division, Abbott GmbH & Co KG, Wiesbaden, Germany*
GI-JA LEE • *Department of Biomedical Engineering, College of Medicine, Kyung Hee University, Seoul, Republic of Korea; Healthcare Industry Research Institute, Kyung Hee University, Seoul, Republic of Korea*
QING LI • *Department of Medicinal Chemistry, University of Minnesota, Minneapolis, MN, USA*
ZHENG LIU • *Department of Chemical Engineering, Tsinghua University, Beijing, China*
ZHIXIA LIU • *Department of Chemical Engineering, Tsinghua University, Beijing, China*
DIANNAN LU • *Department of Chemical Engineering, Tsinghua University, Beijing, China*
GUANGHUI MA • *National Key Laboratory of Biochemical Engineering, Institute of Process Engineering, Chinese Academy of Sciences, Beijing, China*
PAUL MILLNER • *Institute of Membrane and Systems Biology, University of Leeds, Leeds, UK*
MUHAMMAD MONIRUZZAMAN • *Department of Applied Chemistry, Graduate School of Engineering, Kyushu University, Fukuoka, Japan*
TERUYUKI NAGAMUNE • *Department of Bioengineering, Graduate School of Engineering, Center for NanoBio Integration, The University of Tokyo, Tokyo, Japan*
FRANCES NEVILLE • *School of Engineering, University of Newcastle, Callaghan, NSW, Australia*
YAN NIU • *Department of Chemical Engineering, University of Massachusetts Amherst, Amherst, MA; Department of Biochemistry, Purdue University, West Lafayette, IN, USA*
JI HYE PARK • *Department of Biomedical Engineering, College of Medicine, Kyung Hee University, Seoul, Republic of Korea; Healthcare Industry Research Institute, Kyung Hee University, Seoul, Republic of Korea*
HUN-KUK PARK • *Department of Biomedical Engineering, College of Medicine, Kyung Hee University, Seoul, Republic of Korea; Healthcare Industry Research Institute, Kyung Hee University, Seoul, Republic of Korea; Program of Medical Engineering, Kyung Hee University, Seoul, Republic of Korea*
PETER H. PFROMM • *Department of Chemical Engineering, Kansas State University, Manhattan, KS, USA; Institute of Biotechnology 2, Jülich, Germany*
ERWIN REISNER • *Department of Chemistry, University of Cambridge, Lensfield Road, Cambridge CB2 1EW, UK*
MARY E. REZAC • *Department of Chemical Engineering, Kansas State University, Manhattan, KS, USA*
DANIEL J. SAYUT • *Department of Chemical Engineering, University of Massachusetts Amherst, Amherst, MA, USA*
LIANHONG SUN • *Department of Chemical Engineering, University of Massachusetts Amherst, Amherst, MA, USA*
CARSTON R. WAGNER • *Department of Medicinal Chemistry, University of Minnesota, Minneapolis, MN, USA*
LIANG WANG • *School of Chemical & Biomedical Engineering, Nanyang Technological University, Singapore, Singapore*

PING WANG • *Department of Bioproducts and Biosystems Engineering, Biotechnology Institute, University of Minnesota, St. Paul, MN, USA*
BRIAN R. WHITE • *Department of Medicinal Chemistry, University of Minnesota, Minneapolis, MN, USA*
SONGTAO WU • *Department of Bioproducts and Biosystems Engineering, Biotechnology Institute, University of Minnesota, St. Paul, MN, USA*
KERSTIN WÜRGES • *Department of Chemical Engineering, Kansas State University, Manhattan, KS, USA; Institute of Biotechnology 2, Jülich, Germany*
MING YAN • *Department of Chemical Engineering, Tsinghua University, Beijing, China*
GUANGYU ZHU • *MedImmune Vaccines Inc., Gaithersburg, MD, USA*

Chapter 1

Nanoscale-Engineered Cytochrome P450 System with a Branch Structure

Hidehiko Hirakawa and Teruyuki Nagamune

Abstract

Most of the bacterial cytochrome P450s require two kinds of electron transfer proteins, ferredoxin and ferredoxin reductase, and thus P450s do not show catalytic activity by themselves. A microbial transglutaminase-mediated site-specific cross-linking enables the formation of fusion P450 protein with a branched structure, which is generated from a genetic fusion protein of P450–ferredoxin reductase and ferredoxin, an interactive nanoscale protein structure. This fusion P450 system is self-sufficient due to intramolecular electron transfer, which means the system does not require additional electron-transferring proteins. Because some components of bacterial cytochrome P450 system are interchangeable, this self-sufficient system can be applied to non-natural combination of P450 and electron transfer proteins from different species of bacteria.

Key words: Cytochrome P450, P450cam, putidaredoxin, putidaredoxin reductase, intramolecular electron transfer, transglutaminase, post-translational modification, site-specific cross-linking, branched structure.

1. Introduction

In vivo, most enzymes catalyze complicated reactions by interacting and/or cooperating with other proteins including enzymes, while current development of industrial biocatalysts is focused on single enzymes. Future biocatalysts should consist of multiple proteins/enzymes to take proper advantages of enzymes. Genetic fusion has been widely used to manipulate multiple proteins, generating nanoscale protein complexes. However, most fusion proteins are limited to linking two proteins and it is difficult to obtain fully active forms by genetically fusing more than

P. Wang (ed.), *Nanoscale Biocatalysis*, Methods in Molecular Biology 743,
DOI 10.1007/978-1-61779-132-1_1, © Springer Science+Business Media, LLC 2011

two proteins. This is because terminuses of proteins are generally important for protein folding. Enzymatic post-translational modifications, which are important for proteins to have physiological functions in vivo, can provide a solution to problems of genetic fusion. Some enzymes involved in post-translational modification can catalyze site-specific linking between separately expressed active proteins in branched forms as well as linear forms under mild conditions. Therefore, it can develop new generation biocatalysts, breaking limitations of genetic fusion enzymes.

Cytochrome P450s (P450s) can catalyze a wide variety of oxidation reactions using molecular oxygen as an oxidant and requiring electron transfer from NAD(P)H for their catalytic activity (1). Most of the bacterial P450s are soluble and require two kinds of electron transfer proteins, ferredoxin and ferredoxin reductase. For example, P450 from *Pseudomonas putida* (P450cam), which is the most studied bacterial P450, requires putidaredoxin (Pdx) and putidaredoxin reductase (Pdr) (2). This means that bacterial P450 itself does not show catalytic activity. Due to the requirement of Ferredoxin and Ferredoxin reductase, practical applications of bacterial P450s and their characteristic studies have been limited. Fusion of P450 and electron transfer proteins could generate a self-sufficient P450 that shows catalytic activity itself (3). However, component proteins have to be linked with well-designed or optimized peptide linkers to achieve sufficient activity (4). Although chemical cross-linking of protein targeting reactive functional groups (–SH, $-NH_2$, –COOH) may also generate fusion proteins, chemical cross-linking is not site-specific and it is difficult to obtain homogeneous cross-linked products.

A transglutaminase from *Streptomyces mobaraensis* (TGase), which has been widely used in food industry (5), can catalyze the formation of an ε-(γ-glutamyl) lysine bond between the side chains of a glutamine residue and a lysine residue. TGase can recognize specific amino acid sequences and catalyze site-specific cross-linking of tagged proteins with the recognition sequence under protein-friendly conditions (6–9). TGase-mediated site-specific protein cross-linking enables the formation of a branched fusion P450 from a genetically fused protein of P450–ferredoxin reductase linked with a specific sequence containing a single reactive glutamine residue for TGase and ferredoxin tagged with a specific sequence containing a single reactive lysine residue (10, **Fig. 1.1**). A branched fusion P450 composed of bacterial P450, Pdx, and Pdr is self-sufficient because branched structure simply increases the local concentrations of electron transfer proteins and electrons can be transferred efficiently within a fused molecule. This chapter describes how to prepare and characterize branched P450s (10).

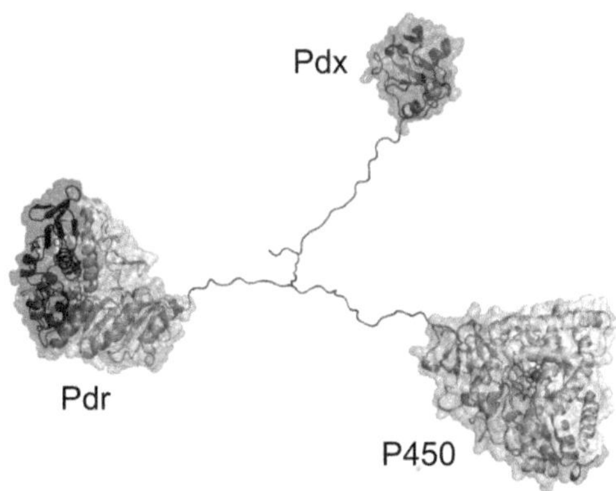

Fig. 1.1. Molecular model of a branched fusion P450.

2. Materials

2.1. Purification of TGase

1. TGase was kindly provided by Ajinomoto Co. Inc. in a semipurified form.
2. Binding buffer A: 50 mM potassium phosphate, pH 7.4, 150 mM KCl.
3. Elution buffer A: 50 mM potassium phosphate, pH 7.4, 50 mM imidazole, 500 mM KCl.
4. Millex-GV syringe-driven filter unit (0.22 μm PDVF membrane filter, Millipore, Billerica, MA, USA).
5. HisTrap FF crude column (1.6×2.5 cm, GE Healthcare, Uppsala, Sweden).
6. Amicon Ultra-15 centrifugal filter device (30,000 NMWL, Millipore).
7. Superdex 75 10/300 GL (10×30 cm, GE Healthcare).
8. BCA Protein Assay Reagent Kit (Pierce Biotechnology, Rockford, IL).

2.2. Protein Expression

1. A fusion protein of Pdr and P450 (Pdr-Qlinker-P450) is constructed by genetically linking the C-terminus of Pdr and the N-terminus of P450 with a reactive glutamine residue containing peptide (Qlinker, GGGGSHEAELRPLAQSHATRHRIPGGGGS, *see* **Note 1**). The enterokinase recognition sequence (DDDDK) is added at the C-terminus of P450 to remove a His_6-tag. A Pdr-Qlinker-P450 encoding gene is cloned into pET22b(+) between *Nde*I and *Not*I sites (**Fig. 1.2a**).

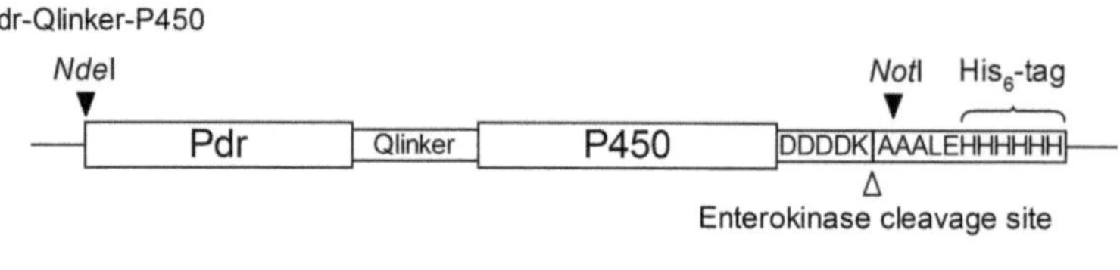

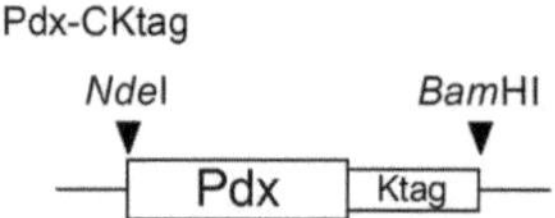

Fig. 1.2. Vector construct of Pdr-Qlinekr-P450 (**a**) and Pdx-CKtag (**b**).

2. A tagged Pdx (Pdx-CKtag) is constructed by genetically adding a reactive lysine residue containing peptide (Ktag, GGGGSLVPRSGSSHHHHHHTSHATRKIPIR, *see* **Note 1**) at the C-terminus of the C73S/C85S mutant of Pdx (*see* **Note 2**). A Pdx-CKtag encoding gene is cloned into pET11a between *Nde*I and *Bam*HI sites (**Fig. 1.2b**).
3. LB medium: 10 g/L Bacto Tryptone (Becton, Dickinson and Company, Sparks, MD, USA), 5 g/L Bacto Yeast Extract (Becton, Dickinson and Company), and 10 g/L NaCl. Adjust at pH 7.4 and autoclave.
4. Terrific broth (TB) medium: Dissolve 12 g of Bacto Tryptone, 24 g of Bacto Yeast Extract, and 4 mL of glycerol in 900 mL of deionized water. Dissolve 12.54 g of K_2HPO_4 and 2.31 g of KH_2PO_4 in 100 mL of deionized water. Autoclave, allow liquids to cool for less than 60°C, and then mix.
5. *Escherichia coli* BL21 Star (DE3) pLysS (Invitrogen, Carlsbad, CA, USA).

2.3. Protein Purification

1. 4-(2-Aminoethyl)benzenesulfonylfluoride (AEBSF, Sigma, St. Louis, MO, USA).
2. Benzonase (Sigma).
3. *d*-Camphor (Wako Pure Chemical Industries, Osaka, Japan).
4. Enterokinase (Invitrogen).
5. Binding buffer B: 50 mM potassium phosphate, pH 7.4, 40 mM imidazole, 500 mM KCl (*see* **Note 3**).
6. Elution buffer B: 50 mM potassium phosphate, pH 7.4, 500 mM imidazole, 500 mM KCl (*see* **Note 3**).
7. EK buffer: 5 mM potassium phosphate, pH 7.4 (*see* **Note 3**).
8. Binding buffer C: 50 mM potassium phosphate, pH 7.4 (*see* **Note 3**).

9. Gel filtration buffer: 50 mM potassium phosphate, pH 7.4, 150 mM KCl (*see* **Note 3**).
10. Binding buffer D: 50 mM potassium phosphate, pH 6.0.
11. HiTrap Desalting (1.6×2.5 cm, GE Healthcare).
12. HiTrap DEAE FF (1.6×2.5 cm, GE Healthcare).
13. HiTrap SP FF (1.6×2.5 cm, GE Healthcare).
14. Superdex 200 10/300 GL (10×30 cm, GE Healthcare).
15. Amicon Ultra-15 Centrifugal Filter Device (10,000 NMWL, Millipore).
16. Amicon Ultra-15 Centrifugal Filter Device (50,000 NMWL, Millipore).

2.4. Preparation of Branched Fusion P450

1. Amicon Ultra-15 Centrifugal Filter Device (100,000 NMWL, Millipore).

2.5. SDS-Polyacrylamide Gel Electrophoresis (SDS-PAGE) Analysis

1. Separating buffer (4X): 1.5 M Tris–HCl, pH 8.7, 0.4% SDS. Store at 4°C.
2. Thirty percent acrylamide/bis solution (29:1 with 3.3% C). Store at 4°C.
3. Stacking gel mixture: Mix 30 mL of 30% acrylamide/bis solution, 25 mL of 1 M Tris–HCl, pH 6.8, 2 mL of 10% (w/v) SDS, and 143 mL of deionized water. Store at 4°C.
4. Glycerol: Prepare 50% solution in water and store at 4°C.
5. Ammonium persulfate (APS, Wako Pure Chemical Industries): Prepare 10% solution in water and store at 4°C.
6. *N*,*N*,*N*,*N*′-Tetramethyl-ethylenediamine (TEMED, Bio-Rad, Hercules, CA, USA).
7. Water-saturated 1-butanol: Vigorously shake equal volumes of water and 1-butanol in a plastic centrifuge tube and allow to separate. Use the top layer. Store at room temperature.
8. Anode buffer: 200 mM Tris–HCl, pH 8.9 (*see* **Note 4**). Store at room temperature.
9. Cathode buffer: 100 mM Tris, 100 mM tricine, 0.1% (w/v) SDS (*see* **Note 4**). Store at room temperature.
10. Laemmli buffer (4X): 25% (v/v) 1 M Tris–HCl, pH 6.8, 20% (v/v) β-mercaptoethanol, 8% (w/v) sodium dodecyl sulfate (SDS), 20% (w/v) sucrose, 0.008% (w/v) bromophenol blue solution in deionized water.
11. Protein molecular weight markers: Broad Range Protein Molecular Weight Markers (Promega, Madison, WI, USA).

12. Staining solution: 0.3% (w/v) Coomassie Brilliant Blue R-250 (Wako Pure Chemical Industries), 10% (v/v) acetic acid, 50% methanol solution in deionized water. Store at room temperature.
13. Destaining solution: 10% (v/v) acetic acid, 30% (v/v) methanol solution in deionized water. Store at room temperature.

2.6. Pyridine Hemochromogen Assay

1. Solution A: Mix 2 mL of pyridine, 1 mL of 1 N NaOH, and 5 mL of deionized water.
2. Saturated sodium dithionate solution in water.

2.7. Measuring UV–Vis Spectrum of Ferrous–CO Complex State

1. Dilution buffer: 50 mM potassium phosphate, pH 7.4, 150 mM KCl.

2.8. Enzyme Assay Using d-Camphor as a Substrate

1. *d*-Camphor: Prepare 1 mM solution in 50 mM potassium phosphate buffer (pH 7.4) containing 150 mM KCl and store at room temperature.
2. Reduced β-nicotinamide adenine dinucleotide (NADH, Sigma): Prepare 10 mM solution in water and store in aliquots at –80°C.

2.9. Enzyme Assay Using Amplex UltraRed as a Substrate

1. NADH: Prepare 0.25 mM solution by diluting 10 mM solution with 50 mM potassium phosphate buffer (pH 7.4).
2. Amplex UltraRed (Invitrogen): Prepare 30 μg/mL solution by diluting 3 mg/mL of DMSO solution with deionized water. Protect from light by wrapping with aluminum foil and store at −20°C.
3. Superoxide dismutase from bovine liver (SOD, Sigma): Prepare 0.2 mg/mL solution in 50 mM potassium phosphate buffer (pH 7.4) and store in aliquots at −80°C.
4. Flexible 96-well plate (flat bottom, without lid; Becton Dickinson Labware, Franklin Lakes, NJ, USA).

3. Methods

P450cam is the most studied P450 because it is easily expressed using *E. coli* and its native substrate, *d*-camphor, and electron transfer partner proteins, Pdr and Pdx, which had been revealed three decades ago (2). A P450 from *Sulfolobus acidocaldarius* (CYP119) is the first discovered thermostable P450 (11).

Although native substrates and electron transfer partner proteins have not been well understood, CYP119 can catalyze a hydroxylation of lauric acid using Pdr and Pdx as electron transfer partner proteins (12). Here, we describe construction and characterization of branched P450 fusion proteins containing P450cam mutant (Q7N/Q211N/Q214N/K215R/Q273N/Q312N/K314R/K314R/Q344N/K345R/Q389N/Q391N/K413R, *see* **Note 5**) or CYP119.

A branched fusion P450 is prepared by stoichiometric cross-linking of Pdr-Qlinker-P450 and Pdx-CKtag, which are separately expressed by *E. coli* and purified. TGase-catalyzed reaction proceeds almost completely, nonetheless, purification steps are required to remove unreacted substrates and TGase. Pdr-Qlinker-P450 and a branched fusion P450 are easily separated by metal chelate affinity chromatography because Pdr-Qlinker-P450 lacks a His_6-tag but a branched fusion P450 contains a His_6-tag derived from Pdx-CKtag. After the removal of unreacted Pdr-Qlinker-P450, a branched fusion P450 can be purified by removing TGase and Pdx-CKtag using gel filtration chromatography.

Spectroscopy analyses are very useful for characterization of a branched fusion P450. The concentration of a branched fusion P450 is accurately determined using pyridine hemochromogen assay and molecular extinction coefficients can be calculated. UV–vis spectrum of a branched fusion P450 in the ferrous–CO complex state indicates the state of heme domain in a branched fusion P450 because an active ferrous–CO complex shows an absorption peak at 450 nm, although an inactive one shows its peak at 420 nm (13). The activity of a branched fusion P450 can be determined from the oxidation rate of NADH by measuring absorbance of NADH at 340 nm. It can also be estimated by fluorescence of deamidated product of Amplex UltraRed (*see* **Note 6**).

3.1. Purification of TGase

1. Add 20 mL of binding buffer A to 2 g of the powder of TGase. Insoluble components are removed by centrifugation at 6,000 × *g* for 30 min.
2. The supernatant is filtrated with a Millex-GV syringe-driven filter unit.
3. The filtrated sample is loaded on a HisTrap FF crude column pre-equilibrated with binding buffer A.
4. The column is washed with 50 mL of binding buffer A, and TGase is eluted with elution buffer A (*see* **Note 7**).
5. TGase-containing fractions are combined and concentrated with an Amicon Ultra-15 Centrifugal Filter Device (30,000 NMWL).

6. The concentrated protein is loaded on a Superdex 75 10/300 GL column pre-equilibrated with binding buffer A. Proteins are eluted with binding buffer A at 0.5 mL/min.
7. The highest purity fractions are combined and concentrated with an Amicon Ultra-15 Centrifugal Filter Device (30,000 NMWL).
8. The concentration of the purified TGase is estimated using BCA Protein Assay Reagent Kit according to the instruction.
9. The purified TGase is stored at –80°C until use.

3.2. Protein Expression

1. *E. coli* BL21 Star (DE3) pLysS is transformed with an expression plasmid.
2. A single colony is inoculated in 5 mL of LB containing 100 μg/mL ampicillin and 34 μg/mL chloramphenicol and cells are grown at 37°C overnight.
3. The grown cells are added to 1 L of TB containing 100 μg/mL ampicillin and 34 μg/mL chloramphenicol and cultivated at 37°C.
4. When OD at 600 nm reaches a value of 0.8, 1 mmol of IPTG and 100 mg of ampicillin (and 1 mmol of ALA for the expression of Pdr-Qlinker-P450) are added and the temperature is lowered at 27°C.
5. After overnight culture, the cells are harvested by centrifugation at 22,000×*g* for 20 min.

3.3. Protein Purification

3.3.1. Purification of Pdr-Qlinker-P450

1. The harvested cells are resuspended with binding buffer B containing 0.1 mM AEBSF and 50 U/mL benzonase and disrupted by 15 repeats of ultrasonication (SONIFIER 250, duty cycle 30%, output control 8) for 1 min with a 3-min interval, with cooling in an ethanol/ice bath.
2. The crude lysate is centrifuged at 22,000×*g* for 30 min and the supernatant is loaded on a HisTrap FF crude column pre-equilibrated with binding buffer B.
3. The column is washed with 50 mL of binding buffer B and Pdr-Q-P450 is eluted with elution buffer B.
4. The colored fractions are loaded on a HiTrap Desalting column and the protein is eluted with EK buffer.
5. The eluted protein is treated with 4 units of enterokinase at 20°C overnight.
6. The cleaved protein is loaded on a HisTrap FF crude column and eluted with EK buffer.
7. The eluted protein is loaded on a HiTrap DEAE FF column and the column is washed with 50 mL of binding buffer C.

8. The protein is eluted from the column with a 0–400 mM KCl gradient. The highest purity fractions (*see* **Note 8**) are combined and concentrated with an Amicon Ultra-15 Centrifugal Filter Device (50,000 NMWL).
9. The concentrated protein is loaded on a Superdex 200 10/300 GL column pre-equilibrated with gel filtration buffer. Proteins are eluted with gel filtration buffer at 0.5 mL/min.
10. The highest purity fractions (*see* **Note 8**) are combined and concentrated with an Amicon Ultra-15 Centrifugal Filter Device (50,000 NMWL).
11. The concentration of the purified protein is determined by pyridine hemochromogen assay (*see* **Section 3.5**).
12. The purified Pdr-Qlinker-P450 is stored at −80°C for long-term storage.

3.3.2. Purification of Pdx-CKtag

1. Pdx-CKtag is partially purified using a HisTrap crude FF column as described above.
2. The colored fractions are loaded on a HiTrap Desalting column and the protein is eluted with binding buffer D.
3. The eluted protein is loaded on a HisTrap SP FF and the column is washed with 50 mL of binding buffer D.
4. The protein is eluted from column with a 0–500 mM KCl gradient. The highest purity fractions, which have the highest ratio of absorption at 412 nm than that at 280 nm, are combined and concentrated with an Amicon Ultra-15 Centrifugal Filter Device (10,000 NMWL).
5. The concentrated protein is loaded on a Superdex 75 10/300 GL column pre-equilibrated with gel filtration buffer. Proteins are eluted with gel filtration buffer at 0.5 mL/min.
6. The highest purity fractions, which have the highest ratio of absorption at 412 nm to that at 280 nm, are combined and concentrated with an Amicon Ultra-15 Centrifugal Filter Device (10,000 NMWL).
7. The concentration of the purified protein is calculated from $\varepsilon_{412\ \mathrm{nm}} = 11.0$ mM/cm (14).
8. The purified Pdx-CKtag is stored at −80°C until use or for long-term storage.

3.4. Preparation of Branched Fusion P450

1. A mixture of 50 μM Pdr-Qlinker-P450 and 50 μM Pdx-CKtag is incubated with 1 μM TGase in binding buffer C at 4°C overnight (*see* **Note 9**).
2. The reaction mixture is loaded on a HisTrap FF crude column pre-equilibrated with binding buffer B.

3. The column is washed with 50 mL of binding buffer B. TGase and unreacted Pdr-Qlinke-P450 are removed in this step.
4. Branched fusion P450 and unreacted Pdx-CKtag are eluted with elution buffer B.
5. Branched fusion P450-containing fractions are combined and concentrated with an Amicon Ultra-15 Centrifugal Filter Device (100,000 NMWL).
6. The concentrated protein is loaded on a Superdex 200 10/300 GL column pre-equilibrated with gel filtration buffer. Proteins are eluted with gel filtration buffer at 0.5 mL/min. Unreacted Pdx-CKtag is removed in this step.
7. The highest purity fractions (*see* **Note 8**) are combined and concentrated with an Amicon Ultra-15 Centrifugal Filter Device (100,000 NMWL).
8. The concentration of the purified protein is determined by pyridine hemochromogen assay (*see* **Section 3.6**).
9. The purified branched fusion P450 is stored at −80°C for long-term storage.

3.5. SDS-PAGE Analysis

1. These instructions assume the use of a Mini-Slab Size Electrophoresis System (Atto, Tokyo, Japan). You can also use other company's minigel electrophoresis systems.
2. Prepare a 1.0 mm thick, 7.5% gel by mixing 1.8 mL of separating buffer (4X), with 1.8 mL acrylamide/bis solution, 2.1 mL water, 1.44 mL glycerol solution, 72 μL of ammonium persulfate solution, and 10 μl TEMED. Pour the gel, leaving space for a stacking gel, and overlay with water-saturated 1-butanol. The gel should polymerize in about 10 min at room temperature.
3. Pour off the 1-butanol and rinse the top of the gel with tap water.
4. Prepare the stacking gel by mixing stacking gel mixture with 20 μl of ammonium persulfate solution and 2 μl TEMED. Pour the stacking gel and insert the comb. The stacking gel should polymerize in about 20 min at room temperature.
5. Add the anode buffer to the lower chamber of the gel unit and set the gel at the gel unit.
6. Pour the cathode buffer into the upper chamber and carefully remove the comb.
7. Mix 3 μl of 5 μM protein sample and 9 μl of Laemmli buffer (4X) and then incubate the mixture at 98°C for 3 min.

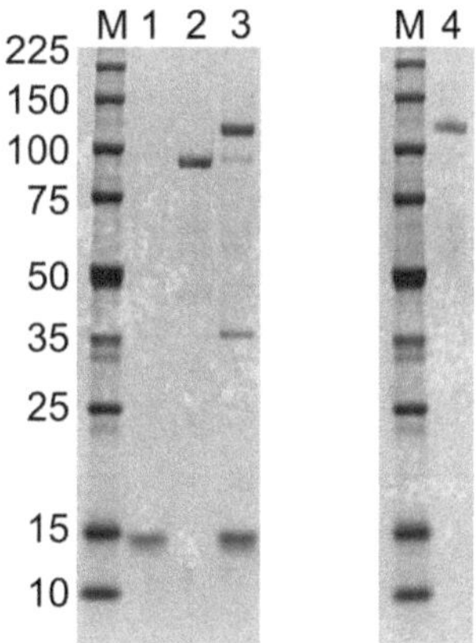

Fig. 1.3. SDS-PAGE analysis for site-specific cross-linking product by TGase (*lane 1*, Pdx-CKtag; *lane 2*, Pdr-Qlinker-P450cam; *lane 3*, the reaction mixture; *lane 4*, purified branched fusion P450cam; excess amount of Pdx-CKtag is added for visualization).

8. Load 2 μl of each sample in a well and include one well for molecular weight markers.
9. Connect to a power supply. The gel is run at 40 mA and electrophoresis is stopped when the dye fronts reach the edge of the gel.
10. Rinse the gel with tap water several times and stain with the staining solution for 3 min.
11. Pour off staining solution and rinse with tap water several times.
12. Pour destaining solution and add a couple of sheets of KimWipe. An example of the result produced is shown in **Fig. 1.3**.

3.6. Pyridine Hemochromogen Assay

1. A dilution series of heme-containing protein sample is prepared. A Soret peak absorbance for each sample is measured.
2. Mixture of 0.6 mL of solution A and 0.3 mL of protein solution is put into 10 mm path length quartz cuvette. The absorptions at 541 nm ($Abs_{541}{}^{ox}$) and 557 nm ($Abs_{557}{}^{ox}$) are measured at 25°C.
3. Ten microliters of saturated sodium dithionate solution is mixed with the above mixture well. The absorptions at 541 nm ($Abs_{541}{}^{red}$) and 557 nm ($Abs_{557}{}^{red}$) are measured at 25°C.
4. Heme concentrations are calculated as follows:

$$[\text{Heme}]\,(\mu\text{M}) = \frac{(Abs_{557}{}^{red} - Abs_{557}{}^{ox}) - (Abs_{541}{}^{red} - Abs_{541}{}^{ox})}{0.0207} \times 3$$

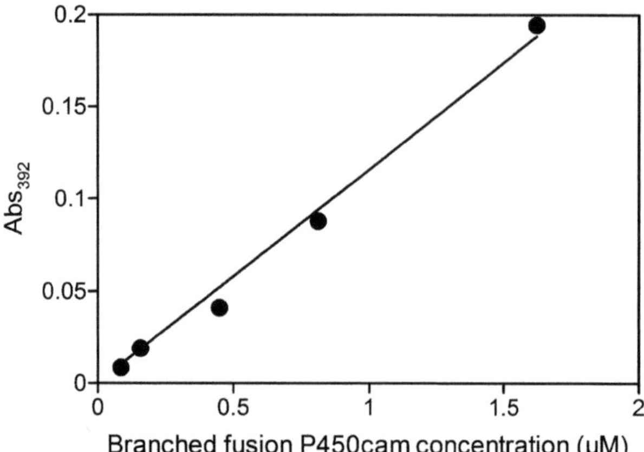

Fig. 1.4. Relationship between absorption of a Soret peak and protein concentration, which is calculated from pyridine hemochromogen assay, of a branched fusion P450cam.

5. The molecular extinction coefficient is determined by plotting a Soret peak absorbance for each protein sample vs. heme concentration. An example of plotting for a branched fusion P450cam is shown in **Fig. 1.4**.

3.7. Measuring UV–Vis Spectrum of Ferrous–CO Complex State

1. Put 0.8 mL of protein solution (~50 μM) in dilution buffer into 10 mm path length quartz cuvette and measure UV–vis spectrum.
2. Add 10 μL of saturated sodium dithionate solution and mix well. Quickly measure UV–vis spectrum.
3. Bubble the reduced protein solution with pure CO gas for 10 s under a fume hood. Quickly measure UV–vis spectrum. Examples of UV–vis spectra of various states are shown in **Fig. 1.5**.

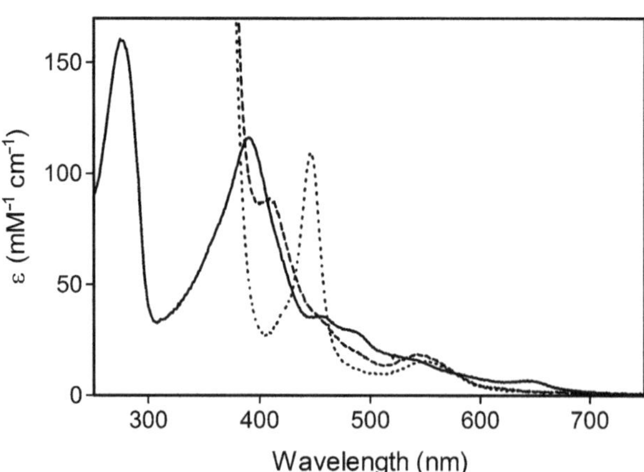

Fig. 1.5. UV–vis spectra of a branched fusion P450cam in ferric (*solid line*), dithionate-reduced (*broken line*), and ferrous–CO complex (*dotted line*) state.

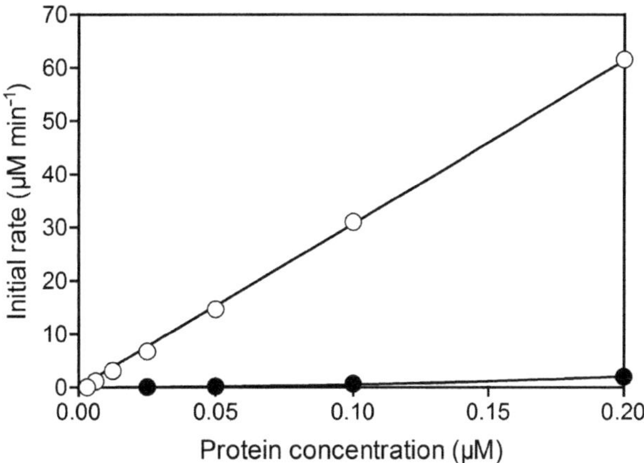

Fig. 1.6. Initial rate as a function of the concentrations of a branched fusion P450cam (*open circles*) and the mutant P450cam reconstituted in a 1:1:1 ration with Pdr and Pdx-CKtag (*closed circles*).

3.8. Enzyme Assay Using d-Camphor as a Substrate

1. Put 1.5 mL of reaction buffer into a standard rectangular quartz cuvette (10 mm path length, 2 mL working volume) equipped with a magnetic stir bar. The reaction mixture is stirred at 400 rpm.
2. Remove 25 μL of reaction buffer with a pipette.
3. Add 500 μL of 1 mM *d*-camphor solution to the cuvette and start recording absorption at 340 nm.
4. Add 20 μL of 10 mM NADH solution to the cuvette.
5. Wait until the absorption at 340 nm reaches a plateau, and then add 5 μL of 20 μM protein solution to the cuvette.
6. Measure the decrease rate of the absorption ($\Delta Abs_{340}/\Delta t$).
7. Reaction rate (V) is calculated as follows:

$$V(\mu M/\min) = \frac{\Delta Abs_{340}/\Delta t(\min^{-1})}{6.22 \times 10^{-3}(\mu M^{-1}\,cm^{-1})}$$

An example of relationship between reaction rate and protein concentration is shown in **Fig. 1.6**.

3.9. Enzyme Assay Using Amplex UltraRed as a Substrate

1. Eighty microliters of buffer is added to each well of flexible 96-well plate.
2. Add 5 μL of 0.2 mg/mL SOD solution (*see* **Note 10**).
3. Add 5 μL of 0.25 mM NADH solution.
4. Add 5 μL of Amplex UltraRed solution.
5. Add 5 μL of protein solution and then quickly start measuring the fluorescent intensity at 590 nm (35 nm bandwidth)

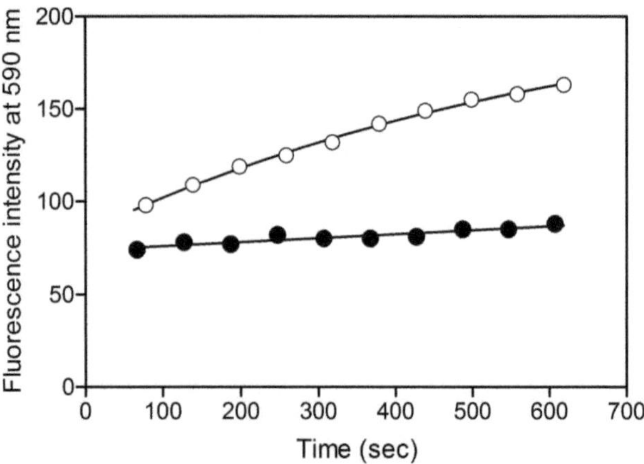

Fig. 1.7. NADH-dependent deamidation of Amplex UltraRed by a branched fusion CYP119 (*open circles*, [NADH] = 12.5 mM; *closed circle*, [NADH] = 0 μM).

excited at 530 nm (25 nm bandwidth) for every 1 min. An example of monitoring deamidation of Amplex UltraRed is shown in **Fig. 1.7**.

4. Notes

1. Several peptide sequences have been reported to act as a substrate for TGase (4–7, 15). We mainly use derivative sequences from the F-helix of horse heart myoglobin (HEAELKPLAQSHATKHKIPIK, reactive residues shown in underline).
2. The wild-type Pdx is highly sensitive to oxygen and easy to denature in aerobic condition. Although the mutation of C73S/C85S in Pdx partially decreases the electron transfer activity, it remarkably improves the stability (14). We usually use this mutant as a component of P450cam/Pdx/Pdr system.
3. Buffers dissolving P450cam and a branched P450 containing P450cam should have 5 mM *d*-camphor for stabilization of P450cam. It takes a long time to dissolve *d*-camphor due to its hydrophobicity. We recommend making it to small pieces before adding to buffer and stirring overnight with a cover.
4. We use Tris–Tricine anode buffer and cathode buffer as running buffer for a wide range of separation.
5. A wild type of P450cam has reactive amino acid residues for TGase, while a mutant P450cam, whose every potentially

exposed glutamine and lysine residues are substituted with asparagine and arginine residues, respectively, does not have reactive residues. The mutant P450cam shows almost same *d*-camphor hydroxylation activity with a wild type (8).

6. Amplex UltraRed is a derivative of Amplex Ultra (10-acetyl-3,7-dihydroxyphenoxazine), which is known as a substrate for P450s (16) as well as horseradish peroxidase.
7. Although TGase does not have a His_6-tag, it weakly binds to Ni–NTA resins.
8. Purity of each fraction is estimated from the ratio (Rz) of the absorbance of a Soret band to that at 280 nm. The higher Rz indicates higher purity of heme-containing protein.
9. The cross-linking reaction is almost complete in 6 h and by-products do not increase at least within 16 h.
10. Amplex UltraRed is sensitive to superoxide. SOD can scavenge superoxide and inhibit background reaction. Ascorbate is known as a scavenger for superoxide and 10 μM of ascorbate can also be used to inhibit background reaction.

Acknowledgments

We are grateful to Ajimonoto Co. Inc. for providing the TGase sample.

References

1. Ortize de Montellano, P. R. (ed.) (1995) *Cytochrome P450: Structure, Mechanisms and Biochemistry.* Plenum Press, New York, NY.
2. Gunsalus, I. C., and Wanger, G. C. (1978) Bacterial P-450cam methylene monooxygenase components: Cytochrome *m*, putidaredoxin, and putidaredoxin reductase. *Methods Enzymol.* **52**, 166–188.
3. Hlavica, P. (2009) Assembly of non-natural electron transfer conduits in the cytochrome P450 system: A critical assessment and update of artificial redox constructs amenable to exploitation in biotechnological areas. *Biotechnol. Adv.* **27**, 103–121.
4. Robin, A., Roberts, G. A., Kisch, J., Sabbadin, F., Grogan, G., Bruce, N., Turner, N. J., and Flitsch, S. L. (2009) Engineering and improvement of the efficiency of a chimeric [P450cam-RhRed reductase domain] enzyme. *Chem. Commun.* **45**, 2478–2480.
5. Yokoyama, K., Nio, N., and Kikuchi, Y. (2004) Properties and applications of microbial transglutaminase. *Appl. Microbiol. Biotechnol.* **64**, 447–454.
6. Kamiya, N., Tanaka, T., Suzuki, T., Takazawa, T., Takeda, S., Watanabe, K., and Nagamune, T. (2003) S-peptide as a potent peptidyl linker for protein cross-linking by microbial transglutaminase from *Streptomyces mobaraensis. Bioconjug. Chem.* **14**, 351–357.
7. Takazawa, T., Kamiya, N., Ueda, H., and Nagamune, T. (2004) Enzymatic labeling of a single chain variable fragment of an antibody with alkaline phosphatase by microbial

tranglutaminase. *Biotechnol. Bioeng.* **86**, 399–404.

8. Tanaka, T., Kamiya, N., and Nagamune, T. (2004) Peptidyl linkers for protein heterodimerization catalyzed by microbial transglutaminase. *Bioconjug. Chem.* **15**, 491–497.
9. Tanaka, T., Kamiya, N., and Nagamune, T. (2005) N-terminal glycine-specific protein conjugation catalyzed by microbial transglutaminase. *FEBS Lett.* **579**, 2092–2096.
10. Hirakawa, H., Kamiya, N., Tanaka, T., and Nagamune, T. (2007) Intramolecular electron transfer in a cytochrome P450cam system with a site-specific branched structure. *Protein Eng. Des. Sel.* **20**, 453–459.
11. McLean, M. A., Maves, S. A., Weiss, K. E., Krepich, S., and Sligar, S. G. (1998) Characterization of a cytochrome P450 from the acidothermophilic archaeon *Sulfolobus solfataricus. Biochem. Biophys. Res. Commun.* **252**, 166–172.
12. Koo, L. S., Immoos, C. E., Cohen, M. S., Farmer, P. J., and Ortize de Montellano, P. R. (2002) Enhanced electron transfer and laurix acid hydroxylation by site-directed mutagenesis of CYP119. *J. Am. Chem. Soc.* **124**, 5684–5691.
13. Martinis, S. A., Blanke, S. R., Hanger, L. P., Sligar, S. G., Hoa, G. H. B., Rux, J. J., and Dawson, J. H. (1996) Probing the heme iron coordination structure of pressure-induced cytochrome P420cam. *Biochemistry* **35**, 14530–14536.
14. Sevrioukova, I. F., Garcia, C., Li, H., Bhaskar, B., and Poulos, T. L. (2003) Crystal structure of putidaredoxin, the [2Fe–2S] component of the P450cam monooxygenase system from *Peudomonas putida. J. Mol. Biol.* **333**, 377–392.
15. Sugimura, Y., Yokoyama, K., Nio, N., Maki, M., and Hitomi, K. (2008) Identification of preferred substrate sequences of microbial transglutaminase from *Streptomyces mobaraensis* using a phage-displayed peptide library. *Arch. Biochem. Biophys.* **477**, 379–383.
16. Rabe, K. S., Kiko, K., and Niemeyer, C. M. (2008) Characterization of the peroxidase activity of CYP119, a thermostable P450 from *Sulfolobus acidocaldarius. ChemBioChem* **9**, 420–425.

Chapter 2

Chemically Induced Self-Assembly of Enzyme Nanorings

Brian R. White, Qing Li, and Carston R. Wagner

Abstract

Continued exploration into the field of chemically induced dimerization (CID) has revealed a number of applications for its use in a broader context as a method of structural assembly (1–4). In particular, the use of CID technology to generate self-assembled (and selectively disassembled) protein toroids serves as a key advancement toward developing stable and controllable protein-based platforms. Such structures have broad application to the development of novel therapeutics, lab-on-a-chip technologies, and multi-enzyme assemblies (5, 6). This chapter describes a method of developing an enzymatically active protein nanostructure incorporating both a CID-based assembly region containing dihydrofolate reductase (DHFR) and an enzymatic region consisting of histidine triad nucleotide binding protein 1 (Hint1). Details of both the production and the characterization of this structure are provided.

Key words: Enzyme nanorings, DHFR, Hint1, nanostructures, self-assembly, chemically induced dimerization, bis-methotrexate, gel filtration, protein expression, protein purification.

1. Introduction

Chemically induced dimerization is the controlled dimerization of proteins via dimerizers. During the process of dimerization, the dimerizers assemble proteins into homospecific or heterospecific multivalent nanostructures. Mimicking the functions of biological inducers for protein dimerization to regulate cellular signaling pathway, CID systems have been developed as investigative tools for the selective activation of various cellular processes to control cell membrane receptor signal transduction and gene expression (7, 8). To demonstrate the competency of this technique, CIDs have also been exploited in the three-hybrid methodology for high-throughput bioscreening. Furthermore, the application of CID as a means to control the assembly of protein oligomers

P. Wang (ed.), *Nanoscale Biocatalysis*, Methods in Molecular Biology 743,
DOI 10.1007/978-1-61779-132-1_2,

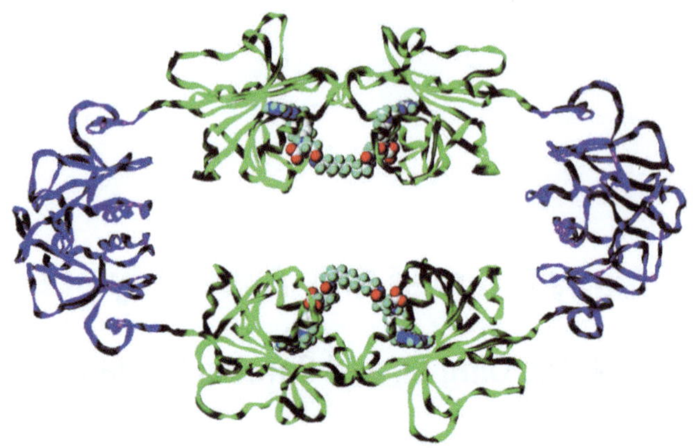

Fig. 2.1. Model of the DHFR–hHint1 dimeric nanoring. DHFR is shown in *green* and Hint1 in *purple*. The dimerizer is rendered in space-filling fashion. Figure reproduced from (5) with permission from the American Chemical Society.

offers an avenue to engineer protein-based materials and nanostructures.

The method of engineering enzyme nanorings described in this chapter is based on the CID system for the preparation of self-assembled protein macrocyclic oligomers of dihydrofolate reductase fusion proteins with a polypeptide linker of variable length ($DHFR_2$) by chemical induction with bis-methotrexate (bis-MTX) (2). The size of the nanorings is dependent on the length of the polypeptide linker between the two DHFRs. Human Hint1 is a highly stable homodimer and acts as a phosphoramidate and acyl-adenylate hydrolase. When incorporating enzyme human Hint1 between the two DHFRs, enzymatically active protein nanorings can be assembled (**Fig. 2.1**) (5). Similar to other $DHFR_2$-based protein nanorings, enzyme nanoring size is also dependent on the length and composition of the polypeptide linking the fusion proteins. The more general assembly of protein nanorings can be characterized by size-exclusion chromatography and their resulting enzymatic activity explored by typical activity assays.

2. Materials

2.1. General

1. Dithiothreitol (DTT) is a common reducing agent that should not be added to buffers until after pH adjustments are made. It will also become oxidized over time – do not use buffers containing DTT that is over ~1 week old.
2. The preparation of bis-methotrexate (**Fig. 2.2**) involves synthetic organic chemistry which is beyond the scope of this

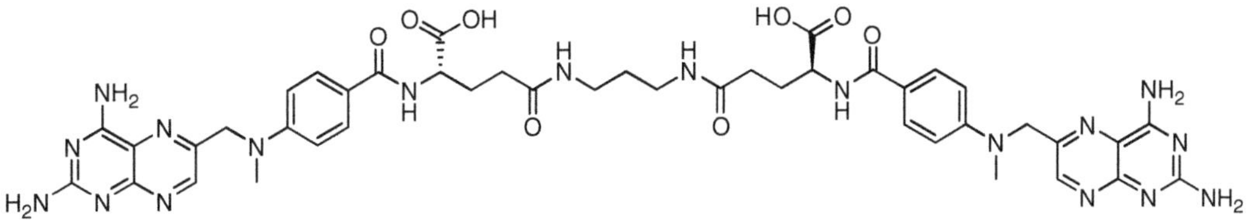

Fig. 2.2. Structure of bis-methotrexate. The preparation of the molecule is detailed in (8).

Fig. 2.3. The DHFR–hHint1 fusion protein. The DHFR proteins serve as the chemically induced dimerization domain while human Hint1 associates naturally and retains its enzymatic activity. The 3DH-GLE nomenclature stands for DHFR–hHint1 with a gly–leu–glu linker between the two proteins. Figure reproduced from (5) with permission from the American Chemical Society.

protocol, but a detailed description of its preparation has been published (9).

3. The construction of the DHFR–human Hint1 (hHint1) plasmid will not be covered here since the construction has been detailed elsewhere (10). This plasmid, or any suitable fusion of an enzyme gene to the DHFR gene, can be transformed into an expression cell line via any number of commercially available protocols (i.e., via Invitrogen- or Promega-competent cell kits).
4. As noted previously, this work utilizes a DHFR–hHint1 fusion protein (**Fig. 2.3**). Hint1 enzymatic activity is measured via the hydrolysis of a fluorogenic substrate, tryptamine 5′-adenosine phosphoramidate (TpAd), and this assay is described in detail elsewhere (11). However, the use of novel DHFR–enzyme constructs is encouraged, and enzymatic activity assays must be tailored to the new DHFR fusion partner.

2.2. DHFR Activity Assay

1. MTEN assay buffer: 50 mM 2-morpholinoethanesulfonic acid (MES), 25 mM Tris(hydroxymethyl)aminomethane–HCl, 25 mM ethanolamine, 100 mM NaCl, pH 7.0, 1 mM DTT. It may be helpful to make a 10× concentrated stock of this buffer and store at 4°C. In this case, add DTT only when diluting to the 1× assay buffer as to prevent premature oxidation of the DTT.
2. Dihydrofolate (DHF) is prepared fresh as described (12) and stored as slurried aliquots under argon at –80°C.

3. Nicotinamide adenine dinucleotide phosphate, reduced form (NADPH), should be dessicated and stored at –20°C.

2.3. Cell Growth and Lysis

1. Luria–Bertani powder (LB) is made by adding 20 g LB powder to 1 L of deionized water and autoclaving at 121°C for 20 min prior to use. LB broth may be stored at 4°C.
2. Ampicillin (Amp) is dissolved in deionized water at a stock concentration of 100 mg/mL, filter sterilized using a 0.22 μm syringe filter, stored at –20°C in aliquots, and added to LB broth and made to a final concentration of 100 μg/mL. Stock aliquots have a shelf life of ~2 weeks at −20°C and should only be thawed and re-frozen once.
3. Isopropyl β-D-1-thiogalactopyranoside (IPTG) is dissolved in deionized water at a stock concentration of 0.5 M, filter sterilized using a 0.22 μm syringe filter, and stored at −20°C. During protein expression, IPTG is added to the culture to a final concentration of 0.5 mM.
4. *E. coli* strain BB2: This strain is produced by disrupting the *E. coli hinT* gene in strain BW25113 as described (13). It is necessary to use hinT-deficient *E. coli* to avoid wild-type *E. coli* Hint1 contamination.
5. Lysis buffer A: 50 mM Tris–HCl, pH 8.0, 5 mM ethylenediaminetetraacetic acid (EDTA), 1.0 mg/mL lysozyme, 50 μg/mL sodium azide, 1 mM DTT.
6. Lysis buffer B: 1.5 M NaCl, 0.1 M $CaCl_2$, 0.1 M $MgCl_2$, 20 μg/mL DNase I, 1 mM phenylmethylsulfonyl fluoride, 1 mM DTT.
7. Dialysis buffer: 20 mM Tris–HCl, pH 7.0, 1 mM EDTA, 1 mM DTT.

2.4. Fusion Protein Purification

1. Methotrexate (MTX) agarose is stored at 4°C and must be protected from light.
2. Buffer A: 20 mM Tris–HCl, pH 7.0, 1 mM EDTA, 1 mM DTT.
3. Bio-Rad protein assay kit.
4. NuPAGE SDS-PAGE apparatus, 4–12% bis-Tris–HCl minigels, 4× NuPAGE loading buffer, 10× NuPAGE reducing agent, NuPAGE antioxidant, NuPAGE 20× MES-SDS running buffer (Invitrogen). Protein gel electrophoresis can be performed using the Invitrogen NuPAGE protocol.
5. Amicon stirred cell equipped with a YM-10 membrane or, alternatively, Amicon Centricon centrifugal filter devices may be used (with 10 kDa cutoff).
6. Diethylaminoethyl-cellulose DE52 (DEAE-DE52) ion exchange media.

7. Glycerol: Add 30 mL glycerol to 70 mL deionized water and autoclave at 121°C for 20 min to yield a sterile 30% solution of glycerol.
8. Trimethoprim.

2.5. Nanoring Assembly and Characterization

1. P500 buffer: 0.5 M NaCl, 50 mM KH_2PO_4, 1 mM EDTA, pH 7.0.
2. Superdex 200 10/300 GL Gel Filtration column (GE Healthcare) equipped on a Beckman System Gold HPLC with detection at 280 and 302 nm detecting the presence of protein and methotrexate, respectively.
3. Bis-methotrexate (bis-MTX or MTX_2C9) can be prepared as described above. It should be stored at –20°C and protected from light. Given these storage conditions, it is stable for greater than 1 year.

3. Methods

3.1. General

1. Transformation into the non-competent *E. coli* BB2 cell line can be achieved using the method of Hanahan (14). All purification steps are performed at 4°C unless otherwise noted.

3.2. DHFR Activity Assay

1. Add 1 mL MTEN buffer to an aliquot of DHF. The concentration of this stock solution can be determined by diluting 5 μL of the solution into 1 mL of MTEN and checking the A_{280} of the solution. The concentration (μM) is then equal to the absorbance multiplied by the dilution factor divided by 0.028 μM/L/cm, the extinction coefficient for DHF at 280 nm.
2. Dissolve 1–2 mg NADPH in 1 mL MTEN. The concentration of this stock solution can be determined as with DHF, using 0.0062 μM/L/cm as the extinction coefficient of NADPH at 340 nm.
3. Start the reaction by adding 50 μM DHF to a solution of DHFR (2–20 μL depending on estimated concentration) and 100 μM NADPH in MTEN assay buffer (1 mL final assay volume, *see* **Note 1**).
4. Monitor the conversion of NADPH to NADP+ in a UV spectrophotometer measuring absorbance at 340 nm over time. The slope of the resulting linear plot represents the activity in μmol/mg/min.

3.3. Cell Growth and Lysis

1. Starting from a plate of BB2 *E. coli* freshly transformed with the plasmid of interest, pick a single colony and inoculate 10 mL LB broth containing 100 μg/mL ampicillin. Grow the culture overnight at 37°C with shaking at 250 rpm.
2. Use the 10 mL culture to inoculate 1 L of LB broth containing 100 μg/mL ampicillin in a 2 L Erlenmeyer flask. Grow the culture at 37°C with shaking as before to an OD600 of 0.4 (*see* **Note 2**).
3. Add IPTG to a final concentration of 0.5 mM (*see* **Note 3**).
4. Incubate the culture for an additional 2.5 h at 37°C, then centrifuge the culture at 5,000×*g* for 15 min at 4°C. Cell pellets can be frozen and stored at –80°C for up to 1 week.
5. Resuspend the cell pellet in 2 mL of lysis buffer A per gram of cells and incubate at room temperature for 5 min. Digest DNA by adding 2 mL of lysis buffer B per gram of cells and incubating at room temperature for another 25 min.
6. Sonicate the resulting suspension for 12 rounds of 15 s, keeping the temperature of the lysate lower than 20°C. In this setup, a VibraCell VCX750 with temperature probe is used to automatically monitor the sonication process. Thirty-five percent of the maximum power of the unit is used during each pulse.
7. Centrifuge the lysate at 25,000×*g* for 45 min at 4°C to pellet the cell debris. Dialyze the supernatant overnight at 4°C against 2 L of dialysis buffer.

3.4. Fusion Protein Purification

1. Prepare the affinity column by adding 12.5 mL MTX agarose to a clean column, then equilibrating the media with 40 column volumes of buffer A.
2. Load the protein dialysate onto the MTX column at 1 mL/min.
3. Wash the protein with 40 column volumes of buffer A and 60 column volumes of buffer A containing 1 M NaCl, then elute the protein with 150 μM trimethoprim in buffer A. All wash and elution steps are performed at 1 mL/min and 9 mL fractions are collected.
4. Analyze 10 μL of each fraction via the Bio-Rad protein assay kit and assay fractions containing more than 0.1 mg/mL of protein via SDS-PAGE and the DHFR activity assay (*see* **Section 3.1**). Combine fractions containing the desired protein and remove the trimethoprim via buffer exchange (to buffer A) in a stirred Amicon chamber equipped with a YM-10 membrane. This buffer exchange step can also be utilized to concentrate the protein to ~2–3 mg/mL.

5. Prepare the DEAE column by pouring ~50–75 mL of media slurry into a clean column and equilibrate at 1.0 mL/min with 10 column volumes of buffer A.
6. Load the protein onto the DEAE column at 1.0 mL/min.
7. Elute the protein with a linear gradient between buffer A and 0.5 M NaCl in buffer A over 720 min. Collect 9 mL fractions at 1.0 mL/min and assay every other fraction for A_{280} to determine where the protein elutes. Analyze 10 μL of fractions containing significant absorbance (*see* **Note 4**) with the DHFR activity assay and SDS-PAGE.
8. Pool fractions containing the protein of interest and concentrate to 1–2 mg/mL using an Amicon chamber as before. Protein concentration can be determined using the Bio-Rad protein assay kit protocol.
9. Add glycerol to a final concentration of 15% and store the protein in aliquots at −80°C.

3.5. Nanoring Assembly and Gel Filtration

1. The concentration of bis-MTX can be determined spectrophotometrically by diluting 5 μL bis-MTX in 1 mL 0.1 M NaOH and reading the absorbance at 302 nm. This absorbance is multiplied by the dilution factor and divided by the extinction coefficient of bis-MTX, 0.0474 μM/L/cm.
2. To assemble the protein nanorings, add 1.1 moleq of bis-MTX to a sample of the fusion protein (typically 5–100 μM) in a final volume of 1 mL P500 buffer. Incubate at room temperature for 1 h (*see* **Note 5**).
3. Load 500 μL of the nanoring mixture on to the gel filtration column and elute at 0.5 mL/min with P500 buffer. A sample trace can be found in **Fig. 2.4** (*see* **Note 6**).

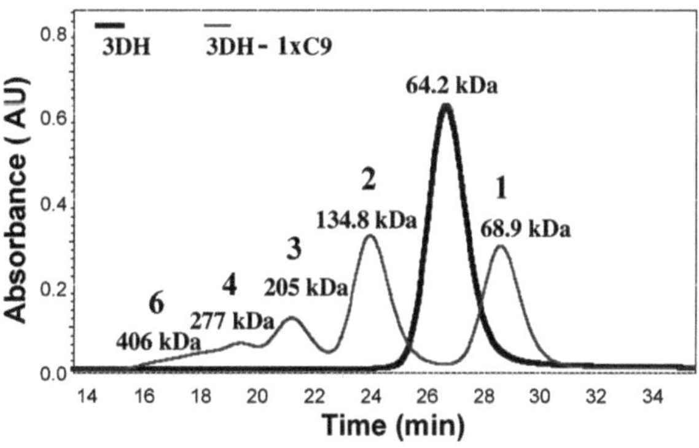

Fig. 2.4. Overlaid size-exclusion profiles for 3DH-GLE (*black*) and 3DH-GLE with 1 eq of bis-MTX added (*red*). *Numbers* above the peaks represent the number of fusion protein monomers (as shown in **Fig. 2.3**) in the ring. The peak eluting at ~29 min represents an intramolecular macrocycle (monomeric ring). Figure reproduced from (5) with permission from the American Chemical Society.

4. Further characterization of the nanorings may be accomplished via transmission electron microscopy, atomic force microscopy, light scattering, or any other desired means.

4. Notes

1. Mixing of reagents in the DHFR activity assay is best achieved by adding the MTEN, NADPH, then protein to a disposable cuvette, then using a piece of parafilm to cap off the vessel and inverting several times. After allowing the solution to incubate at room temperature for at least 1 min to allow NADPH loading onto DHFR, the DHF may be added. The consumption of NADPH begins immediately, so another round of mixing must be performed quickly before all the reagents are depleted. If zero activity is detected in assays where it is thought that the protein concentration is relatively high, it may help to reduce the amount of protein in the sample and try again. Additionally, it is common in samples containing a lot of protein to see the linear reaction rate slow (and become nonlinear) as all the DHF are reduced – velocity measurements should include only the linear region of the UV trace.
2. Induction times and temperatures may vary depending on the fusion protein chosen. Small-scale expression tests on 50 mL cultures are necessary to optimize soluble protein overexpression.
3. Depending on the fusion protein chosen, the concentration of IPTG used for protein induction should be optimized in small-scale protein expression tests before the large-scale protein preparation.
4. Significant absorbance will depend greatly on the robustness of protein expression. When dealing with poorly expressed proteins, checking the ratio of the A_{280} and A_{260} of the peak eluted from the column proves helpful. Ratios tending toward 2.0 are excellent; however, ratios of greater than 1.4 are acceptable. Therefore, if a fraction contains an A_{280} of 0.2 and an A_{260} of 0.1, this would be a good fraction to collect. Conversely, an A_{280} of 0.2 and A_{260} of 0.2 would not. This is generally not an issue with proteins that are expressed very well, as their A_{280}s tend to be much greater (i.e., greater than 2.0).
5. To optimize the oligomerization of the protein nanorings, run the oligomerization reaction in several different ratios of the protein to dimerize. Additionally, the linker length will

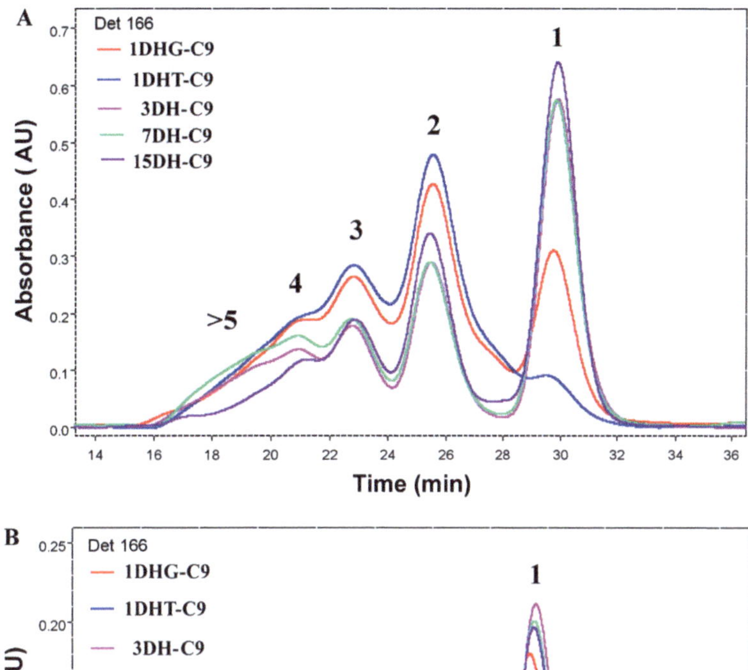

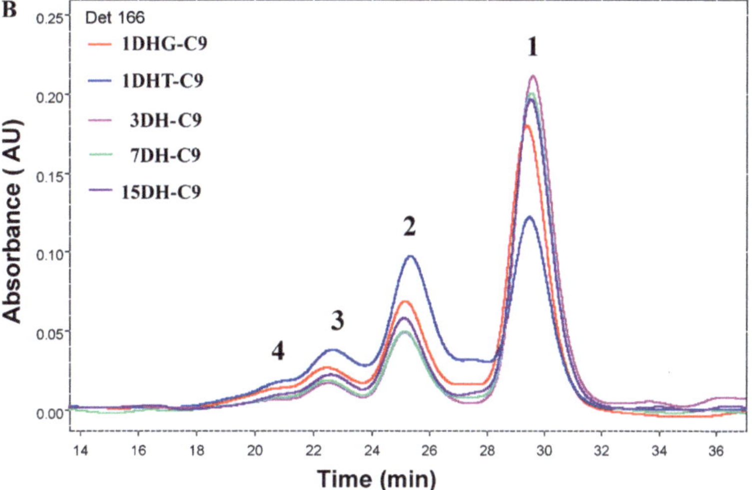

Fig. 2.5. Comparison of bis-MTX polymerized DHFR–hHint1 fusion proteins with different lengths of polypeptide linkers. Overlaid elution profiles for (**a**) 25 μM of protein mixed with 1 eq of bis-MTX and (**b**) 5 μM of protein mixed with 1 eq of bis-MTX. Figure reproduced from (5) with permission from the American Chemical Society. (To follow the absorbance shown above, lines are shown from top to bottom in this order for peak 1 position: 15DH-C9, 7DH-C9, 3DH-C9, 1DHG-C9 and 1DHT-C9).

affect the number of monomers present in the oligomeric ring (**Fig. 2.5**).

6. All the buffers for HPLC-SEC should be filtered through a 0.2 μm filter and degassed before usage. Make sure to equilibrate the column with P500 buffer until the UV baseline is stable before running protein samples. Lastly, after 10 gel filtration runs, wash the column at a flow rate of 0.5 mL/min with 25 mL 0.5 M NaOH, 25 mL deionized water, 25 mL 50% ethanol, 25 mL deionized water, and 50 mL P500 buffer.

References

1. Ballister, E. R., Lai, A. H., Zuckerman, R. N., Cheng, Y., and Mougous, J. D. (2008) In vitro self-assembly of tailorable nanotubes from a simple protein building block. *Proc. Natl. Acad. Sci. USA* **105**, 3733.
2. Carlson, J. C., Jena, S. S., Flenniken, M., Chou, T. F., Siegel, R. A., and Wagner, C. R. (2006) Chemically controlled self-assembly of protein nanorings. *J. Am. Chem. Soc.* **128**, 7630.
3. Dotan, N., Arad, D., Frolow, F., and Freeman, A. (1999) Self-assembly of a tetrahedral lectin into predesigned diamondlike protein crystals. *Angew. Chem. Int. Ed.* **38**, 2363.
4. Ringler, P., and Schulz, G. E. (2003) Self-assembly of proteins into designed networks. *Science* **302**, 106.
5. Chou, T. F., So, C., White, B. R., Carlson, J. C., Sarikaya, M., and Wagner, C. R. (2008) Enzyme nanorings. *ACS Nano* **2**, 2519.
6. Li, Q., Hapka, D., Chen, H., Vallera, D. A., and Wagner, C. R. (2008) Self-assembling antibodies by chemical induction. *Angew. Chem. Int. Ed.* **47**, 10179.
7. Quintarelli, C., Vera, J. F., Savoldo, B., Giordano Attanese, G. M. P., Pule, M., Foster, A. E., Heslop, H. E., Rooney, C. M., Brenner, M. K., and Dotti, G. (2007) Co-expression of cytokine and suicide genes to enhance the activity and safety of tumor-specific cytotoxic t lymphocytes. *Blood* **110**, 2793.
8. Xu, Z. L., Mizuguchi, H., Mayumi, T., and Hayakawa, T. (2003) Regulated gene expression from adenovirus vectors: a systematic comparison of various inducible systems. *Gene* **309**, 145.
9. Carlson, J. C. T., Kanter, A., Thuduppathy, G. R., Cody, V., Pineda, P. E., McIvor, R. S., and Wagner, C. R. (2003) Designing protein dimerizers: the importance of ligand conformational equilibria. *J. Am. Chem. Soc.* **125**, 1501.
10. Chou, T. F., Bieganowski, P., Shilinski, K., Cheng, J., Brenner, C., and Wagner, C. R. (2005) ^{31}P NMR and Genetic analysis establish hinT as the only *E. coli* purine nucleoside phosphoramidase and essential for growth under high salt conditions. *J. Biol. Chem.* **280**, 15356.
11. Chou, T. F., Baraniak, J., Kaczmarek, R., Zhou, X., Cheng, J., Ghosh, B., and Wagner, C. R. (2007) Phosphoramidate pronucleotides: a comparison of the phosphoramidase substrate specificity of human and *Escherichia coli* histidine triad nucleotide binding proteins. *Mol. Pharm.* **4**, 208.
12. Blakley, R. L. (1960) Crystalline dihydropteroylglutamic acid. *Nature* **188**, 231.
13. Datsenko, K. A., and Wanner, B. L. (2000) One-step inactivation of chromosomal genes in Escherichia coli K-12 using PCR products. *Proc. Natl. Acad. Sci. USA* **97**, 6640.
14. Hanahan, D. (1983) Studies on transformation of *Escherichia coli* with plasmids. *J. Mol. Biol.* **166**, 557.

Chapter 3

Self-Assemblies of Polymer–Enzyme Conjugates at Oil–Water Interfaces for Interfacial Biocatalysis

Guangyu Zhu and Ping Wang

Abstract

Many biocatalysts have been shown powerful in enabling reactions among a broad range of substrates possessing very different hydrophilicity/hydrophobicity. Biphasic reaction systems, especially oil–water biphasic systems, have been commonly adopted to mediate such reactions. The greatest challenge in conducting an efficient reaction between two substrates that have to be hosted in two immiscible liquid phases is the mass transfer resistance across interfaces. Imaginably, the substrates afford the most extensive interactions at the interfacial region. The interfacial assembled enzymes, developed by conjugating water-soluble enzymes with hydrophobic polymers, are therefore expected to be efficient in catalyzing biotransformation at the organic–aqueous interfaces. This chapter describes a method in preparing and applying of such interface-assembling enzymes. A model enzyme, α-chymotrypsin (CT), is grafted with polystyrene (PS) to introduce an organic affinity, thus leading to a surfactant-like structure. The characterization of the activity and stability of the interface-assembled enzyme is also presented.

Key words: Biocatalysis, interfacial assembly, polymer–enzyme conjugate, organic–aqueous biphasic system, α-chymotrypsin.

1. Introduction

Organic synthesis is critical to the production of valuable chemicals and pharmaceuticals. In recent years, biocatalysis is rapidly evolving into the synthesis of organic chemicals owing to their remarkably high activity and selectivity, as well as the recent revolutionary advances in discovering and manufacturing new biocatalysts (1–3).

The advances in genetic engineering and protein science have made it feasible to design and generate more active and stable biocatalysts, yet the traditional aqueous biocatalysis has

P. Wang (ed.), *Nanoscale Biocatalysis*, Methods in Molecular Biology 743,
DOI 10.1007/978-1-61779-132-1_3,

substantially limited capability in processing organic chemicals with low aqueous solubility. Over the last three decades or so, noticeable progress has been made in the use of enzymes in nearly anhydrous organic solvents (4–6) and significantly increased catalytic efficiency has been achieved in processing hydrophobic chemicals. However, many important syntheses, such as epoxidation of alkenes (7, 8) and glycosylation of alcohols (9, 10), involve both hydrophobic and hydrophilic substrates and/or cofactors that are insoluble in the same reaction medium and cannot be performed effectively, if not impossible, via monophasic reactions. It has been quite challenging to overcome this solubility barrier in the development of industrial bioprocessing.

In previous efforts dealing with biotransformations that involve immiscible reactants, biphasic reaction systems, which consist of an organic solvent phase to dissolve the hydrophobic chemicals and an aqueous phase to accommodate the enzyme, have been commonly adopted (11, 12). An alternative approach is to use reverse micellar solutions in which enzymes are contained in the micro water pools surrounded by a surfactant layer (13–15). In addition to increasing the substrates' presence in the reactor, the biphasic configuration can also simplify product purification and sometimes can enable reactions that are thermodynamically unfavorable in aqueous solutions (16). However, it has been demonstrated that the partition and diffusion of the substrates and products across the interface strongly limited the overall performance of the biphasic reaction system (17). In addition, it is difficult to reuse the usually expensive enzymes, and the native enzymes dissolved in the aqueous solution are subject to interfacial deactivation upon being adsorbed at oil–water interfaces (18).

In this chapter, a novel interfacial biocatalysis is described, which shows significantly improved efficiency in catalyzing reactions involving both hydrophilic and hydrophobic substrates at the interface of an organic–aqueous biphasic system.

Interfacial self-assembly forms various liquid film structures found in vesicles, cell membranes, and virus–cell fusion. Many biologically important proteins bind to lipid membranes through Coulombic force or hydrophobic affinity. For example, pancreatic lipase reaches the interface via the pancreatic colipase, which provides the necessary hydrophobicity for the interfacial binding of the complex. As another example, hydrophobins, a class of small proteins functioning in the growth and development of fungi, were also found able to self-assemble at O/A interfaces through hydrophobic interactions. Generally speaking, enzymes with relatively large amount of hydrophobic moieties exposed at the surface are less soluble in water but prone to be adsorbed to the air–water or oil–water interfaces, such as lipases and membrane-bound enzymes (19). Most of the industrial enzymes other than lipases, however, generally lack the hydrophobicity to assemble at O/A interfaces, although adsorption of proteins at surfaces can be expected considering thermodynamic driving forces and the amphiphilic nature of the enzymes (20, 21).

Previously developed enzyme modification technologies such as chemical attachment (22), surfactant coating (23), and deglycosylation (24) led to organic-soluble enzymes that were used for homogeneous reactions. To enable interfacial assembly of water-soluble enzymes, it is critical to select hydrophobic moieties that can form a surfactant-like structure once attached to the enzyme surface. To that end, a hydrophobic polymer, polystyrene (PS), is chemically functionalized and covalently attached to a model enzyme, α-chymotrypsin (CT), and the PS–CT conjugate is shown to self-assemble at the interface of an oil–water biphasic system. **Figure 3.1** shows the synthesis of functionalized polystyrene, activation of the polymer, and covalent bond formation between the polymer and the enzyme.

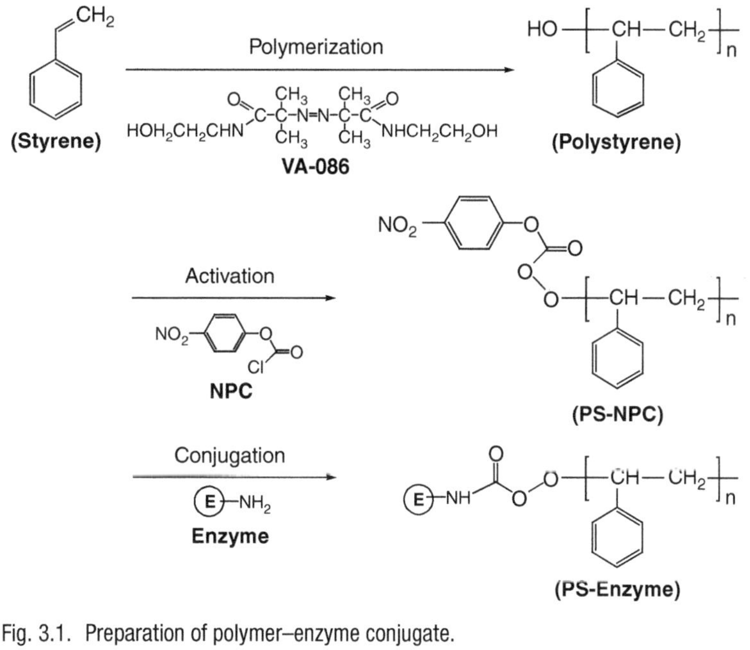

(Styrene) Polymerization $HOH_2CH_2CHN-C(=O)-C(CH_3)_2-N=N-C(CH_3)_2-C(=O)-NHCH_2CH_2OH$ **VA-086** → $HO-[CH-CH_2]_n$ (Polystyrene)

Activation NPC → (PS-NPC)

Conjugation E–NH_2 Enzyme → (PS-Enzyme)

Fig. 3.1. Preparation of polymer–enzyme conjugate.

Interfacial species may undergo three basic patterns of molecular motions: rotation, vibration, and migration (**Fig. 3.2**). Such interfacial motions enable and thus control the interactions among enzymes, substrates, and cofactor at O/A interfaces. Substrates from a bulk phase may approach the enzyme directly or indirectly after being adsorbed to the interface (**Fig. 3.2b**). The interfacial mobility of enzymes will be manipulated by changing the size of modifiers. Specifically, PS of Mw 1,000–6,000,000 Da may be used. Another approach to activity enhancement is by

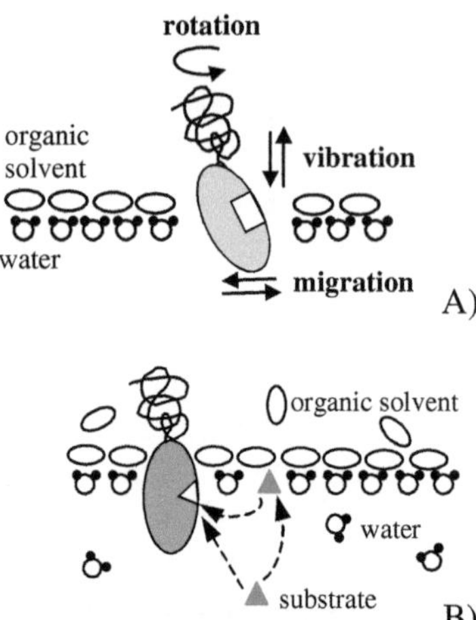

Fig. 3.2. Basic molecular motions and interactions of interface-bound enzymes. (**a**) Interfacial motion of enzyme; (**b**) typical substrate adsorption and interaction at interface.

using emulsion systems which have been studied extensively in seeking increased volume-specific interface area and reduced mass transfer resistance of biphasic reactions. The most common method to achieve emulsion is mechanical stirring, which also improves the interfacial mobility of enzymes.

2. Materials

2.1. Styrene Polymerization

1. Styrene monomer solution in toluene (5.88%, by volume percentage).
2. Initiator 2,2′-azobis [2-methyl-*N*(2-hydroxyethyl) propionamide] (VA-086) solution in *N*,*N*-dimethylformamide (DMF) (62.5 mg/mL).

2.2. Activation of Polystyrene

4-Nitrophenyl chloroformate (NPC) solution in anhydrous methylene chloride (25 mg/mL).

2.3. Conjugation of Enzyme with Activated Polystyrene

1. α-Chymotrypsin (CT) solution in phosphate buffer (3 mg/mL).
2. Phosphate buffer (0.2 M, pH 8.2).

2.4. Protein Assay

1. Biuret reagent A: 15 mg copper sulfate pentahydrate ($CuSO_4{\cdot}5H_2O$) and 45 mg of potassium sodium tartrate are dissolved in 80 mL distilled water followed by adding 2.4 g of sodium hydroxide (NaOH). The solution is diluted to 100 mL with distilled water and kept at room temperature before using.

2. Biuret reagent B: 25 mg ascorbic acid and 37 mg of bathocuproinedisulfonic acid salt are dissolved in distilled water and diluted to 100 mL. Reagents A and B are stable for at least 1 month when stored at room temperature.
3. Bio-Rad protein assay globulin standard solution (1.43 mg/mL in water).

2.5. Active Site Titration

1. 4-Methylumbelliferyl *p*-trimethyl-ammonium cinnamate chloride (4-MUTMAC) solution in phosphate buffer (0.025 mg/mL).
2. Phosphate buffer (0.1 M, pH 7.5).

2.6. Test of the Interfacial Assembly of PS–CT Conjugate

Ten times diluted Bio-Rad protein assay reagent. The absorbance maximum of the Coomassie Brilliant Blue G-250 dye solution shifts from 465 to 595 nm when binding to protein occurs.

2.7. Enzyme Activity Test

1. *n*-Succinyl-ala-ala-pro-phe *p*-nitroanilide (SAAPPN) solution in phosphate buffer (0.5 mM).
2. Phosphate buffer (0.2 M, pH 8.2).

3. Methods

3.1. Styrene Polymerization

1. The VA-086 initiator solution (40 mL) is mixed with the styrene monomer solution (170 mL) in a 500 mL glass bottle and the headspace of the reactor is purged with nitrogen (*see* **Note 1**).
2. The reactor is incubated in a water bath at 85°C for 10 h with magnetic stir at 500 rpm.
3. After reaction, polystyrene is precipitated by adding the reaction mixture drop-wise into 500 mL methanol in a 1 L beaker with strong stir. The precipitated polystyrene is then filtered and further washed with methanol and water to remove residual monomers, solvents, and initiator.
4. The purified polystyrene is dried under vacuum to remove solvents and the molecular weight of polystyrene is measured using gel permeation chromatography (GPC) with a PLgel MIXED-D column.

3.2. Activation of Polystyrene

1. The above-prepared polystyrene (0.5 g) is dissolved in 8 mL toluene and the solution is cooled to 4°C in a refrigerator.
2. 4-Dimethylaminopyridine (DMAP, 12.5 mg) is added into the polystyrene solution.
3. With stir, 2 mL of NPC solution in anhydrous methylene chloride is added slowly into the polystyrene solution.

4. The reaction is proceeded for 5 h at 4°C and then the formed precipitate (4-dimethylaminopyridine hydrochloride) is removed by centrifugation.
5. The NPC-activated polystyrene (PS-NPC) is precipitated by adding the supernatant into 100 mL anhydrous methanol in a beaker with stir. The precipitated PS-NPC is then filtered and further washed with anhydrous methanol to remove NPC and solvent residues (*see* **Note 2**).
6. The purified PS-NPC is dried under vacuum to remove solvents and then used immediately in the next step for the conjugation with enzyme.

3.3. Conjugation of Enzyme with Activated Polystyrene

1. The purified PS-NPC (0.1 g) is dissolved in 3 mL toluene.
2. The PS-NPC toluene solution is added into 3 mL enzyme solution and the mixture is shaken at 4°C for 8 h.
3. After conjugation, the mixture is transferred into Eppendorf tubes (1.5 mL) and centrifuged at 7,826×g for 5 min. The PS–enzyme conjugate partitioned at the interface is collected and washed with toluene (0.5 mL per wash) and buffer (0.5 mL per wash) at least five times to remove unreacted polystyrene and native CT.
4. The purified PS–CT conjugate is vacuum dried and stored at 4°C for future use.

3.4. Protein Assay

The concentration of enzyme solution is measured using the reverse biuret method (25).

1. The enzyme loading of the PS–CT conjugate is measured by adding 1 mg of the conjugate and 10 μL distilled water into 200 μL biuret reagent A. The mixture is incubated at 37°C for 5 min.
2. Biuret reagent B (1 mL) is then added into the mixture and incubation is continued at 37°C for an additional 0.5 min.
3. The solution is filtered through 0.2 μm syringe filter and the absorbance of the filtrate at 485 nm is measured on a UV–Vis spectrophotometer. The protein content of the PS–CT conjugate is calculated according to the calibration curve obtained by following the same procedure except that the 10 μL water is replaced with the standard globulin solutions of different concentrations.

3.5. Active Site Titration

The amount of active α-chymotrypsin in the PS–CT conjugate is measured using a fluorogenic titrant (26).

1. One milligram of PS–CT conjugate is added into 4 mL phosphate buffer containing 0.025 mg/mL 4-MUTMAC.
2. The mixture is shaken at room temperature for 1 min followed by filtration through a 0.2 μm syringe filter.

3. The fluorescence of the filtrate is measured at excitation of 360 nm and emission of 450 nm, and the amount of active CT is calculated according to the standard calibration curve obtained by using active CT at different concentrations.

3.6. Test the Interfacial Assembly of PS–CT Conjugate

1. Two milligrams of PS–CT conjugate is added into a biphasic system containing 6 mL toluene and 6 mL deionized water.
2. As a comparison, 6 mg of native CT is added into a separated glass vial containing the same biphasic system.
3. After stirring for 5 min and phase separation, 1 mL of Bio-Rad protein assay reagent solution is added to each of the biphasic systems.
4. The location of the enzyme is indicated by the color change as shown in **Fig. 3.3**. The blue color is observed at the oil–water interface for PS–CT conjugate, indicating a very selective interfacial assembly. In contrast, native CT remains in the aqueous phase.

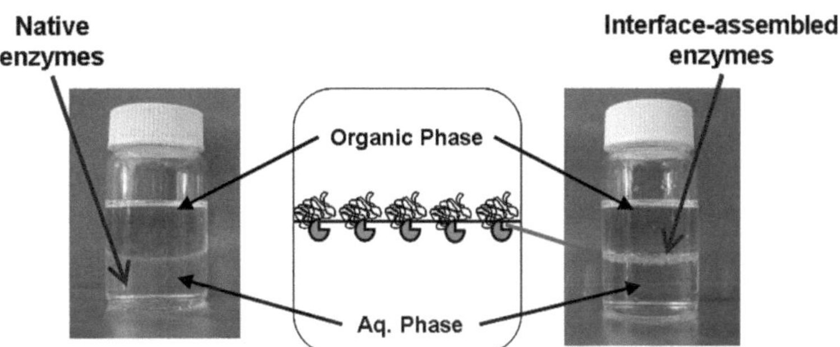

Fig. 3.3. Interfacial assembly of PS–CT conjugate.

3.7. The Activity of PS–CT Conjugate

The activity of the α-chymotrypsin is tested using *n*-succinyl-ala-ala-pro-phe *p*-nitroanilide (SAAPPN) as the substrate (27).

1. For native CT, 10–100 μL of enzyme solution (1 mg/mL in 0.2 M, pH 8.2, phosphate buffer) is added into 2.5 mL buffer containing 0.5 mM of SAAPPN. The reaction rate is measured on a UV–Vis spectrophotometer at 410 nm.
2. For the PS–CT conjugate, 1 mg of the conjugate is added into a biphasic system containing 2.5 mL toluene and 2.5 mL buffer with 0.5 mM SAAPPN.

Periodically, sample is taken from the buffer solution, and the absorbance at 410 nm is measured on a UV–Vis spectrophotometer. By comparing the reaction rate of the PS–CT conjugate to that of the native CT, the relative activity of the conjugate is obtained (*see* **Note 3**).

4. Notes

1. Styrene polymerization is a free radical reaction. Because oxygen scavenges free radicals, it is critical to remove oxygen from the reactor in order for the polymerization to proceed.
2. PS-NPC is an active ester and easily hydrolyzes in aqueous solutions. During the preparation of PS-NPC, water should be avoided to prevent hydrolysis.
3. Typically, an enzyme loading of 53 mg CT/g PS–CT conjugate is obtained and the specific activity of the PS–CT conjugate is 11.7% relative to the native CT. The stability test of the PS–CT conjugate conducted in the toluene–aqueous biphasic system shows a 2.9-fold increase of the half-life time of the enzyme activity compared to that of native CT in the same biphasic system. The polystyrene-conjugated biocatalysts prepared via the technique described in this chapter show improved catalytic activity and enhanced stability compared to their native counterparts. For example, the transgalactosylation reaction involving lactose and hexanol in a toluene–buffer biphasic system catalyzed by interface-assembled polystyrene-β-galactosidase (PS-GL) is 145-fold faster than the same reaction catalyzed by native β-galactosidase. The half-life time of the PS-GL is 4.2-fold longer than that of the native enzyme exposed to the same oil–water biphasic system (28). Similarly, polystyrene-conjugated chloroperoxidase (PS–CPO) shows a 2.5-fold enhancement of enzymatic productivity compared to native CPO in catalyzing the epoxidation of styrene in a styrene–buffer biphasic system with batch-mode reaction; while a 25-fold improvement is realized in a continuous feeding-mode reaction (29). The suppression of side hydrolysis reactions (such as the hydrolysis of lactose in the aqueous phase by β-galactosidase) and the mitigation of enzyme deactivation by the substrates (such as the deactivation of chloroperoxidase by hydrogen peroxide present in the aqueous phase) substantially contribute to the improved overall catalytic efficiency of the interface-assembled PS–enzyme conjugates.

Acknowledgments

This work is supported by grants from the National Science Foundation (CTS-0214769) and the American Chemical Society PRF Program (36726-G4).

References

1. Turner, M. K. (1995) Biocatalysis in organic chemistry (Part II): Present and future. *Trends Biotechnol.* **13**, 253–258.
2. Koeller, K. M., and Wong, C. H. (2001) Enzymes for chemical synthesis. *Nature* **409**, 232–240.
3. Loughlin, W. A. (2000) Biotransformation in organic synthesis. *Bioresour. Technol.* **74**, 49–62.
4. Klibanov, A. M. (2001) Improving enzymes by using them in organic solvents. *Nature* **409**, 246.
5. Blinkovsky, A. M., Martin, B. D., and Dordick, J. S. (1992) Enzymology in monophasic organic media. *Curr. Opin. Biotechnol.* **3**, 124–129.
6. Halling, P. J. (2000) Biocatalysis in low-water media: Understanding effects of reaction conditions. *Curr. Opin. Chem. Biol.* **4**, 74–80.
7. Colonna, S., Gaggero, N., Richelmi, C., and Pasta, P. (1999) Recent biotechnological developments in the use of peroxidases. *Trends Biotechnol.* **17**, 163–168.
8. Palucki, M., Pospisil, P. J., Zhang, W., and Jacobsen, E. N. (1994) Highly enantioselective, low-temperature epoxidation of styrene. *J. Am. Chem. Soc.* **116**, 9333–9334.
9. Ooi, Y., Hashimoto, T., Mitsuo, N., and Satoh, T. (1985) Enzymic formation of β-alkyl glycosides by β-galactosidase from *Aspergillus oryzae* and its application to the synthesis of chemically unstable cardiac glycosides. *Chem. Pharm. Bull.* **33**, 1808–1814.
10. Okahata, Y., and Mori, T. (1998) Transglycosylation catalyzed by a lipid-coated β-D-galactosidase in a two-phase aqueous-organic system. *J. Mol. Catal. B: Enzym.* **5**, 119–123.
11. Antonini, E., Carrea, G., and Cremonesi, P. (1981) Enzyme catalysed reactions in water – Organic solvent two-phase systems. *Enzyme Microb. Technol.* **3**, 291–296.
12. Hickel, A., Radke, C. J., and Blanch, H. W. (1999) Hydroxynitrile lyase at the diisopropyl ether/water interface: Evidence for interfacial enzyme activity. *Biotechnol. Bioeng.* **65**, 425–436.
13. Pier Luigi, L. (1985) Enzymes hosted in reverse micelles in hydrocarbon solution. *Angew. Chem. Int. Ed.* **24**, 439–450.
14. Khmelnitsky, Y. L., Gladilin, A. K., Roubailo, V. L., Martinek, K., and Levashov, A. V. M. S. (1992) Reversed micelles of polymeric surfactants in nonpolar organic solvents. A new microheterogeneous medium for enzymatic reactions. *Eur. J. Biochem.* **206**, 737–745.
15. Miyake, Y. (1996) Enzyme reaction in water-in-oil microemulsions. *Colloids Surf. A* **109**, 255–262.
16. Halling, P. J. (1987) Biocatalysis in multiphase reaction mixtures containing organic liquids. *Biotechnol. Adv.* **5**, 47–84.
17. Chae, H. J., and Yoo, Y. J. (1997) Mathematical analysis of an enzymatic reaction in an aqueous/organic two-phase system: Tyrosinase-catalysed hydroxylation of phenol. *J. Chem. Tech. Biotechnol.* **70**, 163–170.
18. Ross, A. C., Bell, G., and Halling, P. J. (2000) Organic solvent functional group effect on enzyme inactivation by the interfacial mechanism. *J. Mol. Catal. B: Enzym.* **8**, 183–192.
19. Verger, R. (1980) Enzyme kinetics of lipolysis. In *Methods in Enzymology* (Purich, D. L., ed.), Academic Press, Orlando, FL, pp. 340–392.
20. Yampolskaya, G., and Platikanov, D. (2006) Proteins at fluid interfaces: Adsorption layers and thin liquid films. *Adv. Colloid Interface Sci.* **128–130**, 159–183.
21. Milthorpe, B. K. (2005) Protein adsorption to surfaces and interfaces. In *Surfaces and Interfaces for Biomaterials* (Vadgama, P., ed.), Woodhead, Cambridge, pp. 763–781.
22. Wang, P., Woodward, C. A., and Kaufman, E. N. (1999) Poly(ethylene glycol)-modified ligninase enhances pentachlorophenol biodegradation in water-solvent mixtures. *Biotechnol. Bioeng.* **64**, 290–297.
23. Vikram, M. P., and Dordick, J. S. (1994) Mechanism of extraction of chymotrypsin into isooctane at very low concentrations of aerosol OT in the absence of reversed micelles. *Biotechnol. Bioeng.* **43**, 529–540.
24. Vazquez-Duhalt, R., Fedorak, P. M., and Westlake, D. W. S. (1992) Role of enzyme hydrophobicity in biocatalysis in organic solvents. *Enzyme Microb. Technol.* **14**, 837–841.
25. Matsushita, M., Irino, T., Komoda, T., and Sakagishi, Y. (1993) Determination of proteins by a reverse biuret method combined with the copper-bathocuproine chelate reaction. *Clin. Chim. Acta* **216**, 103–111.
26. Gabel, D. (1974) Active site titration of immobilized chymotrypsin with a fluorogenic reagent. *FEBS Lett.* **49**, 280–281.
27. Delmar, E. G., Largman, C., Brodrick, J. W., and Geokas, M. C. (1979) A sensitive new substrate for chymotrypsin. *Anal. Biochem.* **99**, 316–320.

28. Zhu, G., and Wang, P. (2004) Polymer–enzyme conjugates can self-assemble at oil/water interfaces and effect interfacial biotransformations. *J Am. Chem. Soc.* **126**, 11132–11133.

29. Zhu, G., and Wang, P. (2005) Novel interface-binding chloroperoxidase for interfacial epoxidation of styrene. *J. Biotechnol.* **117**, 195–202.

Chapter 4

Molecular Assembly-Assisted Biocatalytic Reactions in Ionic Liquids

Muhammad Moniruzzaman and Masahiro Goto

Abstract

Room temperature ionic liquids (RTILs), having no measurable vapor pressure, represent an interesting class of tunable designer solvents. Due to their many unique properties, ILs have been used as attractive alternatives to environmentally harmful ordinary organic solvents in a wide range of applications including enzymatic biotransformation. Compared to conventional organic solvents, ILs offer many advantages for biocatalysis such as enhanced conversion rates, high enantioselectivity, better enzyme stability, and improved catalyst recoverability and recyclability. However, biocatalysis in ILs has not yet fully achieved its potential because many biocatalysts are insoluble in most ILs. This limitation could be overcome by the formation of nano/micrometer-sized aqueous microemulsion droplets in an IL continuous phase (referred to as water-in-IL microemulsions) stabilized by a layer of surfactants. Enzymes can be dissolved in such water droplets and protected from the unfavorable effect of ILs by the surfactant layer. In this chapter, a simple and effective method for the development of aqueous microemulsion droplets in a hydrophobic IL comprising an anionic surfactant sodium bis(2-ethyl-1-hexyl) sulfosuccinate (AOT) is presented. For this approach, we have synthesized a hydrophobic IL [C_8mim][Tf_2N] (1-octyl-3-methyl imidazolium bis(trifluoromethyl sulfonyl) amide) containing a long pendant hydrocarbon chain to facilitate the dissolution of AOT molecules. A detailed description of the procedure for the potential use of this newly developed water-in-IL reverse microemulsion for biocatalysis is also included.

Key words: Ionic liquids, enzymatic biocatalysis, lipase, horseradish peroxidase, AOT, biotransformation, microemulsions, water droplets, 1-hexanol, green chemistry.

1. Introduction

Ionic liquids (ILs) have been used extensively as alternatives to toxic, hazardous, flammable, and highly volatile organic solvents (1–4). Indeed, their many unique and attractive physicochemical properties such as negligible vapor pressure (4), multiple solvation

P. Wang (ed.), *Nanoscale Biocatalysis*, Methods in Molecular Biology 743,
DOI 10.1007/978-1-61779-132-1_4,

interactions with organic and inorganic compounds (5), excellent chemical and thermal stability, and a wide liquid temperature range make ILs great candidates for volatile organic compound (VOC) replacements. In addition, as a designer solvent, the properties such as the viscosity, hydrophobicity, density, and solubility of ILs can be tuned by selecting different combinations of cations and anions, to customize RTILs for some specific demands. Typical ILs are composed of an organic cation (most often *N,N′*-substituted imidazolium or *N*-substituted pyridinium or tetraalkylated ammonium or tetraalkylated phosphonium) and an inorganic anion (*see* **Fig. 4.1**). Based on the solubility of ILs in water, ILs can be divided into two categories: hydrophobic and hydrophilic. This water miscibility mainly depends on the anions of ILs. In the past decade, ILs have been exploited for use in a wide range of applications, including extraction, organic synthesis, separation, nanomaterial synthesis, and enzymatic reactions. In particular, the application of environmentally benign ILs as solvents or (co)solvents in biocatalytic reactions and processes has received tremendous attention in the past few years (6–13).

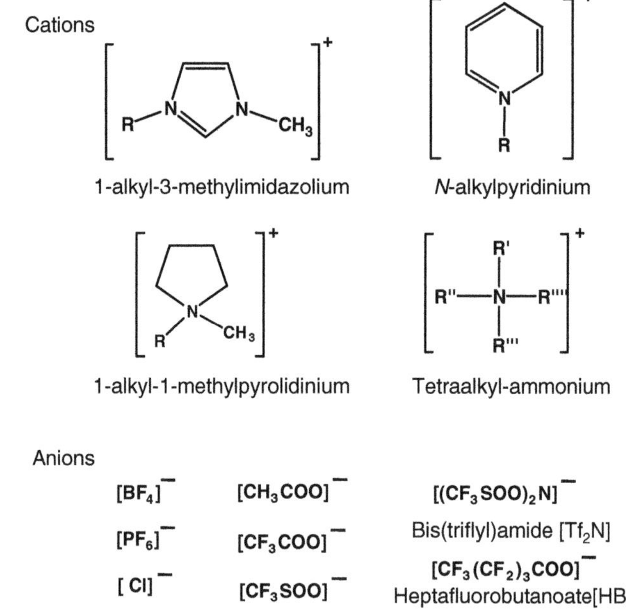

Fig. 4.1. Structures of typical ions of ionic liquids.

The technological utility of enzymes can be enhanced greatly by their use in ILs rather than in conventional organic solvents or in natural aqueous reaction media due to ILs' unique physicochemical properties. In fact, the use of ILs has solved many problems encountered in biocatalytic transformations in aqueous and organic solvents. The first successful report on enzyme-catalyzed reaction in ILs media was published in 2000 by Russell and co-workers (10). Since then, a wide number of enzymes have been subjected to ILs to test their catalytic activity; examples are lipases, proteases, oxidoreductases, peroxidases, and whole cells (6, 7). Compared to those observed in organic solvents, enzymes have presented enhanced stability and selectivity as well as better recoverability and recyclability in many cases. Interestingly, ILs have been shown to act as liquid supports for immobilizing enzyme molecules, providing a suitable microenvironment for biocatalysis in nonaqueous conditions (12). In addition, biocatalysis with an ILs medium may have added advantages as reaction media for carrying out biotransformations with polar or hydrophilic substrates, such as amino acids (10) and carbohydrates (13), which are insoluble or sparingly soluble in most organic solvents (e.g., isooctane and hexane) generally used for nonaqueous biotransformation.

In general, enzymes do not dissolve readily in most ILs. Although some ILs with coordinating anions (e.g., chloride, acetate, nitrate, and dicyanamide), which are strong hydrogen bond acceptor, can dissolve enzymes through weak hydrogen bonding interactions, they often induce enzyme conformational changes resulting in inactivation. For example, EAN (ethylammonium nitrate) can dissolve CaLB (*Candida antarctica* lipase B) through a strongly coordinating nitrate anion resulting in denaturation (7). When free enzymes become active in ILs, they remain suspended as a powder in pure ILs in monophasic or biphasic systems. An effective strategy for enhancing enzyme solubility in ILs involves addition of a small amount of water. As in the case of nonaqueous solvents, the role of water in enzyme-catalyzed reactions in ILs can be analyzed by the same rules as for organic solvents (14). However, the dissolved enzymes show low catalytic activity due to their conformational change in ILs (7, 10). To improve enzyme solubility as well as activity in ILs, modified enzymes (e.g., physically or chemically immobilized enzymes, cross-linked enzyme aggregates (CLEAs), and organic solvent treated enzymes) have been developed to catalyze chemical transformation (e.g., esterification, hydrolysis, and alcoholysis) (6, 7). However, the procedures for the preparation of such modified enzymes are rather complicated. To take the full advantage of the nonvolatile nature of ILs as media for biocatalysis, a superior IL-based solvent system which can dissolve a useful amount of biocatalysts without loss of their catalytic activity is highly desirable.

One of the most promising approaches to solve this problem is to form a colloidal water solution in an IL continuous phase (noted as water-in-ionic liquids (w/IL) microemulsions). Enzymes can be solubilized in the dispersed water phase, just like conventional water-in-oil (w/oil) microemulsions or reverse micelles. It is well documented that enzymes can be solubilized in organic solvents by the use of surfactants without the loss of their catalytic activity (15–17). In this microheterogeneous medium, enzyme molecules are entrapped in tiny water domains and thus become protected against unfavorable contact with the surrounding organic solvent by a layer of water and surfactant molecules, thereby exhibiting good stability and activity. In this chapter, we will focus on two aspects as follows: first, we will describe a simple procedure for the formation of water-in-ionic liquid reverse microemulsions; second, we will present protocols to determine the activity of two model enzymes microencapsulated in such new systems.

The ability to design IL-philic surfactants for the interface between water and hydrophobic ILs offers new opportunities for biocatalysis in nonaqueous environments. In the past years, considerable effort has been devoted to developing thermodynamically stable water droplets in an IL continuous phase (18, 19). However, one important drawback regarding the formation of w/IL microemulsions is that most of conventional surfactants are not soluble in ILs. In fact, the properties of ILs are much more different from those of nonpolar organic solvents. The polarities of water-immiscible ILs (generally referred as hydrophobic IL) are close to those of methanol and acetonitrile (13). To date, very few lipophilic surfactants are found to be soluble in hydrophobic ILs. The formation of bicontinuous microemulsions in mixtures of IL and water stabilized by nonionic surfactants, such as Tween-20 and Trinton X-100, has been reported by Gao and co-workers (18, 20). They hypothesized that dissolution of such surfactants in ILs is attributed to the interaction between the electronegative oxygen atoms of oxyethylene (OE) units of Tween-20 and Trinton X-100 and the electropositive imidazolium ring. However, to form such w/IL microemulsions, a large amount of surfactant is required to solubilize a significant amount of water, which makes the system unsuitable for enzymatic transformation. Recently, we have found that anionic surfactant AOT (sodium bis(2-ethyl-1-hexyl) sulfosuccinate), which has been used extensively to form "organic solvent–water–surfactant" microemulsions in the absence of cosurfactants (21), can be dissolved in hydrophobic ILs (i.e., $[C_8mim][Tf_2N]$ (1-octyl-3-methyl imidazolium bis(trifluoromethyl sulfonyl) amide)) (*see* **Fig. 4.2**) with the aid of 1-alcohols such as 1-hexanol. The addition of 1-hexanol in viscous IL decreased the viscosity significantly, which obviously facilitates the microemulsion formation. Furthermore, the use of 1-hexanol can enhance the formation of micelles by acting as a cosurfactant (22). Upon addition of a small amount of water into AOT–IL–hexanol, solutions form optically transparent aqueous nano-environments in IL continuous phase (19, 23, 24). Here, water may act as a "glue" to bond the surfactant polar head groups together because the hydrophilic SO^{-3} group of AOT molecules can strongly interact with water through the stronger hydrogen bonding (25). Dynamic light scattering (DLS) study

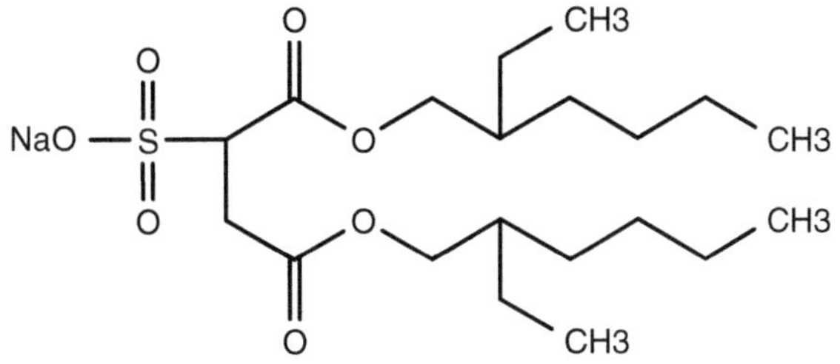

Surfactant: AOT: (sodium *bis*(2-ethyl-1-hexyl) sulfosuccinate)

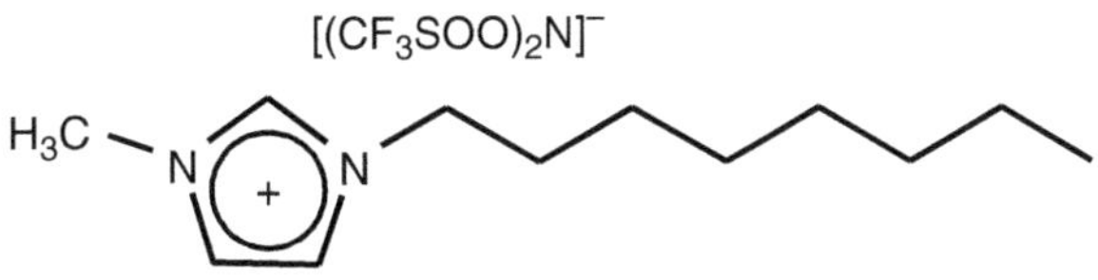

IL $[C_8mim][Tf_2N]$: (1-octyl-3-methylimidazolium *bis* (trifluromethylsulfonyl) amide)

Fig. 4.2. Structure of surfactant and ionic liquid used for the formation of microemulsions.

indicated that increasing the water content of the microemulsions resulted in an increased micelle volume, a common phenomenon for "classic" water-in-oil (w/o) microemulsions (26). These systems yield a spherical structure for AOT aggregates (19), which are similar to (w/o) microemulsions (27). However, the sizes of water pools in IL are somewhat higher and very sensitive to the increment of water, and the variation extent is larger than that for microemulsions formed in organic solvents. These features are probably attributed to the unique structure and properties of ILs. It was revealed that such microemulsions can dissolve various enzymes and other biomolecules in their water core without loss of their catalytic activity (23, 24).

The catalytic activity of lipase PS and horseradish peroxidase, two interfacial active enzymes, microencapsulated in IL-assisted microemulsions was found to be higher than that in microemulsions of AOT with isooctane (23, 24). This enhanced catalytic activity can be explained as follows: first, the partition of the substrates, products, or other molecules involved in the reaction between the aqueous pseudophase and the IL continuous phase; second, a change in the enzyme microenvironment; and third, the presence of 1-hexanol in the reaction medium. As mentioned previously, ILs can dissolve many organic solvents used for organic synthesis, which are poorly soluble in conventional hydrophobic organic solvents (10, 13). For example, the substrate (pyrogallol) and product (purpurogallin) involved in HRP-catalyzed reaction are significantly soluble in IL $[C_8mim][Tf_2N]$ whereas they are poorly soluble in isooctane. Thus, it is reasonable to assume that the product concentration in the aqueous pseudophase was reduced significantly due to the favorable partitioning to the IL continuous phase. Consequently, the product inhibitory effect may be less effective in the IL systems than that in the organic solvents systems. Consequently, this remarkable property may play a significant role in accelerating the HRP oxidation of pyrogallol in w/IL microemulsions. Like water-in-oil microemulsions/micelles, the enzymes' performance in IL-based microemulsions was dependent on the various physicochemical parameters of the system, such as substrate concentrations, W_0 (molar ratio of water to surfactant), pH, and 1-hexanol content (24). The curve of HRP activity–W_0 profile was found to be hyperbolic for the w/IL microemulsions, behavior consistent with "classic" w/o microemulsions (28).

In conclusion, since w/IL microemulsion provides a large interfacial area, two interfacially active enzymes (e.g., lipase PS and HRP) were used as model biocatalysts. On the other hand, IL [C_8mim][Tf_2N] that is relatively more hydrophobic and less viscous was found to be very effective in solubilization of anionic surfactant AOT to form stable water domain in IL continuous phase. In short, we confidently believe that this new microemulsion will offer new opportunities for biotransformation in "green" solvents ionic liquids.

2. Materials

2.1. Synthesis of Ionic Liquid [C_8mim] [Tf_2N] (see Note 1)

1. 1-Methylimidazole.
2. Chloro-octane.
3. Bis-trifluoro methanesulfonide lithium salt [$LiN(CF_3SO_2)_2$] (Sigma-Aldrich, St. Louis, MO).
4. Diethyl ether.
5. Oil bath.

2.2. Preparation of Microemulsions

1. AOT, sodium bis(2-ethyl-1-hexyl) sulfosuccinate (purity >99%, Sigma-Aldrich) (*see* **Fig. 4.2**).
2. 1-Hexanol.
3. Milli Q water.
4. 0.2 μm millipore Millex-LG filter.
5. Karl-Fischer moisture titrator (Mitsubishi, CA-07 Moisturemeter, Japan).

2.3. Lipase-Catalyzed Reaction

1. *Burkholderia cepacia* lipase (former *Pseudomonas cepacia* lipase) (Amano Enzyme, Nagoya, Japan).

2. *p*-Nitrophenyl butyrate (*p*-NPB).
3. 50 mM Tris–HCl buffer, adjusted to pH 8.0.
4. UV–Vis spectrophotometer (Jasco V-570, Japan).

2.4. HRP-Catalyzed Reaction

1. Horseradish peroxidase (HRP) (grade I) (Wako, Japan) (*see* **Note 2**).
2. Pyrogallol (1,2,3-benzenetriol) (Tokyo Chemical Industries, Japan).
3. Extra pure hydrogen peroxide (H_2O_2) (30% w/w) from Sigma. H_2O_2 solutions were prepared by diluting the initial concentrated solution with buffer immediately prior to the experiment.
4. Acetone.
5. 50 mM sodium phosphate buffer, pH 7.0.

3. Methods

3.1. Synthesis of IL

IL [C_8mim][Tf_2N] is prepared by modification of the procedures used for synthesis of various ILs (29, 30). For synthesis of [C_8mim][Tf_2N], first, [C_8mim][Cl] is synthesized and then anion is exchanged as follows.

1. To prepare [C_8mim][Cl], add equal molar amounts of chloro-octane and 1-methylimidazole to a round-bottomed flask fitted with a reflux condenser. Stir the mixture at 60°C for 48 h. After cooling the resulting viscous liquid to room temperature, let the mixture settle to form two phases.
2. After decanting the top phase containing unreacted starting materials, add diethyl ether (a volume approximately equal to half that of the bottom phase) and mix thoroughly. Following the decantation of top phase, add fresh diethyl ether with vigorous mixing and repeat this step five times. Washing with diethyl ether should suffice to remove any unreacted material from the bottom phase.
3. After the last step of removing diethyl ether, remove any remaining diethyl ether by heating the bottom phase to 70°C and stirring while on a vacuum line. Finally, lyophilize the resultant solution to get white powder product of [C_8mim][Cl].
4. To prepare [C_8mim][Tf_2N], transfer 50 g of [C_8mim][Cl] to a 200 mL glass container and mix 25 mL water. Then add Li(Tf_2N) in a molar ratio of 1.1:1 slowly to minimize

the amount of heat generated and mix with constant stirring at room temperature until the formation of two phases. The bottom phase and the top phase contained [C_8mim][Tf_2N] and aqueous LiCl, respectively.

5. After decanting the top phase, add 50 mL of fresh deionized water and mix the mixtures thoroughly. Repeat this step until the washings are totally Cl^- free (*see* **Note 3**).
6. As the last step, remove the remaining water by vacuum followed by lyophilization (*see* **Notes 4** and **5**) and store IL (light yellow color) in the desiccators until use. The purity of IL should be verified by elementary analysis and NMR study.

3.2. Preparation of Microemulsions

1. First, dissolve the required amount of AOT in IL containing 10% (v/v) 1-hexanol (*see* **Notes 6** and 7) to obtain 200 mM AOT in solution.
2. Then, add a small amount of water to prepare a microemulsion (*see* **Note 8**). Clear and stable solutions were obtained by vortex mixing and used as stock solutions for enzymatic reactions.
3. Adjust the molar ratio of water to AOT of the microemulsions (denoted as W_0) by injecting an appropriate amount of buffer solution (*see* **Note 9**).
4. Store in the desiccators until use (19).

3.3. Determination of Lipase Activity

The standard activity test was performed using lipase-catalyzed hydrolysis of *p*-nitrophenyl butyrate as a model reaction (*see* **Fig. 4.3**).

1. Dissolve the enzyme in 50 mM Tris–HCl buffer (pH 8.0) at a concentration of 2 mg/mL and store at 4°C until use.
2. Inject 10 μL enzyme solution into a 0.79 mL microemulsion prepared as described in **Section 3.2** and shake mildly for approximately 30 s to obtain a clear and optically transparent solution.

Fig. 4.3. Lipase-catalyzed hydrolysis of *p*-nitrophenyl butyrate.

3. Commence the reaction by adding 200 μL of *p*-NPB solution (25 mM, prepared in stock microemulsions). The final concentration of substrate was 0.005 M (*see* **Note 10**).
4. Monitor reaction progress at 410 nm (λ_{max} for the product, *p*-nitrophenol) for 5 min using a UV–Vis spectrophotometer (*see* **Note 11**). The reaction temperature was controlled at 35 ± 0.2°C (*see* **Note 12**).
5. From the slopes of the linear portions of the absorbance versus time curves, obtained by least squares fitting, detect the lipase activity as the initial rates (*v*) in μmol/min considering the value of the molar extinction coefficient (Mol ext. = 1,080 M/cm) (*see* **Note 13**) (23).

3.4. Determination of HRP Activity

HRP-catalyzed oxidation of pyrogallol by H_2O_2 was used as a model reaction (*see* **Fig. 4.4**), which has previously been performed in w/o microemulsions.

1. Add 2.5 μL pyrogallol solution (0.8 M in acetone) to 0.97 mL of stock microemulsion with specific amount of buffer solution to reach the desired W_0. Vortex the mixture to obtain a macroscopically homogeneous solution.
2. After adding 5 μL HRP (1.25 mg/mL), shake the contents mildly and keep at 35°C for few minutes, so that they are combined with the reaction mixture at ambient temperature.
3. Finally initiate the reaction by adding 2 μL of H_2O_2 buffer solution (0.4 M).
4. Record the progress of the reaction at 420 nm (λ_{max} for the product, purpurogallin) after immediate addition of hydrogen peroxide.
5. From the slopes of the linear portions of the absorbance versus time curves (linear, at least during the first few minutes), obtained by least squares fitting, the values of the initial rates (*v*) can be measured in μM/min, considering the value of the molar extinction coefficient of purpurogallin in the IL microemulsion (Mol ext. = 3,450 M/cm) (*see* **Note 14**) (24).

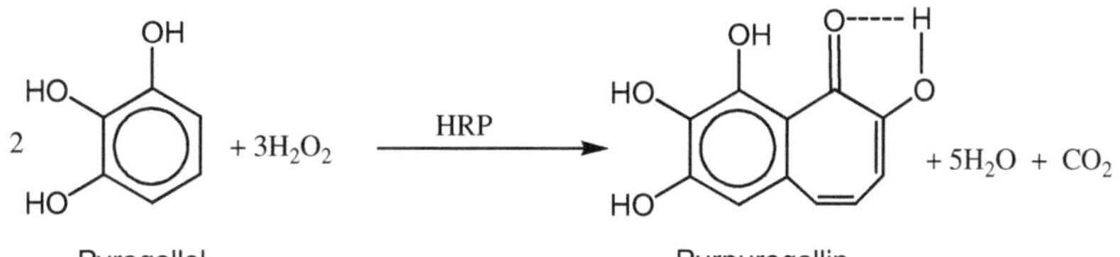

$$2\ \text{Pyrogallol} + 3H_2O_2 \xrightarrow{\text{HRP}} \text{Purpurogallin} + 5H_2O + CO_2$$

Fig. 4.4. Oxidation of pyrogallol by horseradish peroxidase and hydrogen peroxide. The formation of an *orange color* purpurogallin can be monitored by UV–Vis spectrophotometry at $\lambda = 420$ nm.

4. Notes

1. The dissolution of AOT was dependent on the selection of ILs. We have used IL [C_8mim][Tf_2N] because it is relatively more hydrophobic and less viscous. Both of these factors favor the formation of microemulsions. The solubility of AOT was found to be highly dependent on the hydrophobicity of the hydrocarbon chain on the imidazole ring (19), which generally increases with an increase in the alkyl chain length for a fixed anionic group (30).
2. The peroxidase concentration was determined spectrophotometrically using a molar absorption of 9.1×10^4 M/cm at 403 nm (31).
3. IL [C_8mim][Tf_2N] should be Cl^- free because chloride impurities in ILs have the adverse effect on enzyme activity (32). To check any LiCl in the upper aqueous phase, $AgNO_3$ should be used. In addition, one should keep in mind that the purity of an IL is a key factor for their successful application. For example, residual volatiles can have detrimental effects on the performance of the enzymes.
4. Sometimes IL [C_8mim][Tf_2N] becomes brown color. This probably occurs due to existence of oxidation products, thermal degradation products, halides, and so on. In this case, IL should be treated with activated carbon and/or by washing with pure solvents to remove the color as well as other impurities.
5. The final water content in IL is measured by a Karl-Fischer moisture titrator. Due to the high viscosity, dry methanol should be used to dilute the IL.
6. To dissolve 200 mM AOT in IL, minimum 10 vol.% of 1-hexanol is required. Higher concentration can be used to facilitate the dissolution of AOT in the IL. However, 1-alcohols are inhibitors of many enzymes. For example, the enzymatic activity of HRP encapsulated in w/IL microemulsions decreased with the increase of 1-hexanol content (24).
7. The solubility of AOT in IL-hexanol mixture can be accelerated by subjecting to higher temperature or applying ultrasound.
8. Buffer solutions should be used instead of water when using the system for enzymatic reactions. In addition, types of buffer, their composition, and pH should be selected based on the enzymes.

9. Since a trace amount of water can be dissolved in the IL, the W_0 value has been calculated from equation [**1**]:

$$W_0 = \frac{(m_{\mathrm{w}} - m_{\mathrm{w,0}})/M_{\mathrm{w}}}{m_{\mathrm{AOT}}/M_{\mathrm{AOT}}} \qquad [1]$$

where m_{w} is the total mass of water as determined by a Karl Fischer moisture titrator, $m_{\mathrm{w,0}}$ is the mass of water dissolved in the solvent at the experimental temperature, M_{w} stands for the molar mass of water, m_{AOT} denotes the mass of the surfactant, and M_{AOT} is the molecular weight of AOT molecule. Our DLS measurements showed that only after saturating the IL with water, additional water was involved to form water pools (19).

10. All concentrations were given with respect to the total volume of the system to avoid the complexity of the volume fraction of the water droplet in the w/IL microemulsions and partitioning coefficient of substrates in the bulk IL and water pool.
11. A spin bar should be used in 1-cm-long quartz cuvette for proper mixing of the reaction solution.
12. A temperature of 35°C was used to facilitate the handling and transfer of the viscous ionic liquid.
13. All initial rates were corrected by subtracting the nonenzymatic-catalyzed rates of hydrolysis (blank reaction).
14. The blank reaction at pH 7.0 was negligible.

Acknowledgments

We gratefully acknowledge the JSPS (Japan Society for the Promotion of Science) for a JSPS Postdoctoral Fellowship (M. Moniruzzaman) and the necessary funding for this work. The authors would like to thank Prof. N. Kamiya for very helpful discussions.

References

1. Wasserscheid, P., and Welton, T. (ed.) (2003) *Ionic Liquids in Synthesis*. Wiley-VCH, Weinheim.
2. Rogers, R. D., and Seddon, K. R. (ed.) (2002) *Ionic Liquids: Industrial Applications for Green Chemistry*. ACS Symposium Series, Vol. 818. American Chemical Society, Washington, DC.
3. Welton, T. (1999) Room temperature ionic liquids. Solvents for synthesis and catalysis. *Chem. Rev.* **99**, 2071–2084.

4. Earle, M. J., Esperanca, J. M. S. S., Gilea, M. A., Lopes, J. N. C., Rebelo, L. P. N., Magee, J. W., Seddon, J. A., and Widegren, J. A. (2006) The distillation and volatility of ionic liquids. *Nature* **439**, 831–834.
5. Anderson, J. L., Ding, J., Welton, T., and Armstrong, D. W. (2002) Characterizing ionic liquids on the basis of multiple solvation interactions. *J. Am. Chem. Soc.* **124**, 14247–14254.
6. van Rantwijk, F., and Sheldon, R. A. (2007) Biocatalysis in ionic liquids. *Chem. Rev.* **107**, 2757–2785.
7. Yang, Z., and Pan, W. (2005) Ionic liquids: Green solvents for nonaqueous biocatalysis. *Enzyme Microb. Technol.* **37**, 19–28.
8. Turner, M. B., Spear, S. K., Huddleston, J. G., Holbrey, J. D., and Rogers, R. D. (2003) Ionic liquid salt-induced inactivation and unfolding of cellulase from *Trichoderma reesei*. *Green Chem.* **5**, 443–447.
9. Kragl, U., Eckstein, M., and Kaftzik, N. (2002) Enzyme catalysis in ionic liquids. *Curr. Opin. Biotechnol.* **13**, 565–571.
10. Erbeldinger, M., Mesiano, A. J., and Russel, A. J. (2000) Enzymatic catalysis of z-aspartame in ionic liquid: An alternative to enzymatic catalysis in organic solvents. *Biotechnol. Prog.* **16**, 1129–1131.
11. Nakashima, K., Maruyama, T., Kamiya, N., and Goto, M. (2005) Comb-shaped poly (ethylene glycol)-modified subtilisin Carlsberg is soluble and highly active in ionic liquids. *Chem. Commun.* 4297–4299.
12. Lozano, P., de Diego, T., and Iborra, J. L. (2006) Immobilization of enzymes for use in ionic liquids. In *Methods in Biotechnology* (Jones, J. M., ed.), Humana, Totowa, NJ, pp. 257–268.
13. Park, S., and Kazlauskas, R. J. (2001) Improved preparation and use of room temperature ionic liquids in lipase-catalyzed enantio- and region-selective acylation. *J. Org. Chem.* **66**, 8395–8401.
14. Carrea, G. (1984) Biocatalysis in water-organic solvent two-phase systems. *Trends Biotechnol.* **2**, 102–106.
15. Martinek, K., Levashov, A. V., Khmelnitskii, Y. L., Klyachko, N. L., and Berezin, I. V. (1982) Colloidal solution of water in organic solvents: A microheterogenous medium for enzymatic reactions. *Science* **218**, 889–891.
16. Luisi, P. L., and Magid, I. J. (1986) Solubilization of enzymes and nucleic acids in hydrocarbon micellar solutions. *CRC Crit. Rev. Biochem.* **20**, 409–474.
17. Ayala, A. G., Kamat, S., Beckman, E. J., and Russell, A. J. (1992) Protein extraction and activity in reverse micelles of a nonionic surfactant. *Biotechnol. Bioeng.* **29**, 806–814.
18. Gao, Y., Li, N., Zheng, L., Zhao, X., Zhang, S., Han, B., Hou, W., and Li, G. (2006) A cyclic voltammetric technique for the detection of micro-regions of bmimPF_6/Tween-20/H_2O microemulsions and their performance characterization by UV-vis spectroscopy. *Green Chem.* **8**, 43–49.
19. Moniruzzaman, M., Kamiya, N., Nakashima, K., and Goto, M. (2008) Formation of reverse micelles in a room temperature ionic liquid. *Chem. Phys. Chem.* **9**, 689–692.
20. Gao, Y., Han, B., Han, B. X., Li, G., Shen, D., Li, Z., Du, J., Hou, W., and Zhang, G. (2005) TX-100/water/1-butyl-3-methylimidazolium hexafluorophosphate micro-emulsions. *Langmuir* **21**, 5681–5684.
21. De, T., and Maitra, A. (1995) Solution behavior of Aerosol OT in non-polar solvents. *Adv. Colloid Interface Sci.* **59**, 95–193.
22. Moulik, S. P., Digout, L. G., Aylward, W. M., and Palepu, R. (2000) Studies on the interfacial composition and thermodynamic properties of W/O microemulsions. *Langmuir* **16**, 3101–3106.
23. Moniruzzaman, M., Kamiya, N., Nakashima, K., and Goto, M. (2008) Water-in-ionic liquid microemulsions as a new medium for enzymatic reactions. *Green Chem.* **10**, 497–500.
24. Moniruzzaman, M., Kamiya, N., and Goto, M. (2009) Biocatalysis in water-in-ionic liquid microemulsions: A case study with horseradish peroxidase. *Langmuir* **25**, 977–982.
25. Ritter, R. E., Undiks, U. P., and Levinger, N. E. (1998) Impact of counterion on water motion in Aerosol-OT reverse micelles. *J. Am. Chem. Soc.* **120**, 6062–6067.
26. Zulauf, M., and Eicke, H. F. (1979) Inverted micelles and microemulsions in ternary systems H_2O/Aerosol-OT/isooctane as studied by photon correlation spectroscopy. *J. Phys. Chem.* **83**, 480–486.
27. Laia, C. A. T., López-Cornejo, P., Costa, S. M. B., d'Oliveira, J., and Martinho, J. M. G. (1998) Dynamic light scattering study of AOT microemulsions with nonaqueous polar additives in an oil continuous phase. *Langmuir* **14**, 3531–3537.
28. Azevedo, M., Fonseca, L. P., Graham, D. L., Cabral, J. M. S., and Prazeres, D. M. F. (2001) Behavior of horseradish peroxidase in AOT reverse micelles. *Biocatal. Biotransform.* **19**, 213–233.

29. Bonhôte, P., Dias, A. P., Papageorgiou, N., Kalyanasundaram, K., and Grätzel, M. (1996) Hydrophobic, highly conductive ambient-temperature molten salts. *Inorg. Chem.* **35**, 1168–1178.
30. Huddleston, J. G., Visser, A. E., Reichert, W. M., Willauer, H. D., Broker, G. A., and Rogers, R. D. (2001) Characterization and comparison of hydrophilic and hydrophobic room temperature ionic liquids incorporating the imidazolium cation. *Green Chem.* **3**, 156–164.
31. Meathly, A. C. (1955) Plant peroxidase. *Methods Enzymol.* **2**, 801–813.
32. Lee, S. H., Sung, S. H., Lee, S. B., and Koo, Y. M. (2006) Adverse effect of chloride impurities on lipase-catalyzed transesterifications in ionic liquids. *Biotechnol. Lett.* **28**, 1335–1339.

Chapter 5

Organic-Soluble Enzyme Nano-Complexes Formed by Ion-Pairing with Surfactants

Songtao Wu, Andreas Buthe, and Ping Wang

Abstract

The solubilization of enzymes in organic solvents for non-aqueous biocatalysis has attracted considerable attention since the homogeneous distribution accounts for a drastically improved reaction efficiency compared to enzymes dispersed as aggregates in an organic phase. This chapter highlights ion-pairing as a valuable and facile method to make enzymes soluble in organic solvents. Ion-pairing denotes the formation of a nano-complex, in which a single enzyme molecule in the core is surrounded by counter-charged surfactant molecules. The special architecture of this nano-complex exposes the surfactant hydrophobic group toward the bulk solvent and renders the complex sufficiently soluble in organic media. This chapter also describes the underlying principle of ion-pairing as well as simple preparation and characterization techniques to yield highly active enzyme–surfactant nano-complexes. The general applicability of this technique is demonstrated on the base of the hydrolytic enzyme α-chymotrypsin (α-CT) and the redox enzyme glucose oxidase (GO_x).

Key words: Enzyme solubilization, hydrophobic ion-pairing, organic-soluble enzyme, enzyme–surfactant nano-complex, α-chymotrypsin, glucose oxidase, cationic surfactant, anionic surfactant.

1. Introduction

Over the last three decades, it has been found that enzymes, originally evolved to function in an aqueous environment, can also function highly selectively in non-aqueous environments for transformations that are thermodynamically unfavorable in aqueous media (1). Meanwhile some substrates and/or products may undergo undesired side reactions in an aqueous environment, so they are only sufficiently stable in organic media (2). Consequently, the utilization of enzymes in organic solvents became attractive and resulted in a vast number of novel and

P. Wang (ed.), *Nanoscale Biocatalysis*, Methods in Molecular Biology 743,
DOI 10.1007/978-1-61779-132-1_5, © Springer Science+Business Media, LLC 2011

interesting enzyme-catalyzed biotransformations, such as stereoselective transesterification (3–6) and esterification (7, 8), regioselective acylation of glycols and sugars (9, 10), and regiospecific oxidation of phenols (11). The direct suspension of lyophilized enzyme powders in the monophasic organic phase is the most obvious way to deploy a biocatalyst in this way, but due to the inherent properties of enzyme proteins, dispersion on a molecular level cannot be achieved and aggregates are formed (12). Even though the size of the enzyme aggregates can be controlled to a certain extent, mass transfer limitations will not allow all enzyme molecules to contribute equally to the overall reaction. As suggested in literatures one remedy for this is the immobilization of the enzyme molecules by binding onto or by entrapment into hydrophobic materials (13). In doing so the enzyme molecules are much better dispersed and easily accessible for the substrate compared to anhydrous enzyme aggregates. Moreover their separability from the reaction media and hence their reusability is facilitated (13). Intensive research over the last years has culminated in a number of highly sophisticated approaches to immobilize enzymes, e.g., in form of enzyme–polymer composites (14). Some of those composites can be even applied beyond classical biocatalysis. For instance, "bioplastics" in the form of thin films or coatings are of particular interest for a vast number of applications such as in the realm of biosensing or antifouling surfaces (14, 15). Despite the great potential, the entrapment of enzymes in common hydrophobic "plastic" polymers like polystyrene, poly(methyl methacrylate), and poly(vinyl acetate) might encounter the same above-mentioned problems in terms of dispersion and aggregation, which greatly impairs the performance of bioactive composite materials. For the preparation of those composites it is highly beneficial to dissolve the enzyme molecule in the same phase as the hydrophobic polymers in order to achieve a homogeneous distribution.

A real homogeneous dispersion on a molecular level can only be achieved if the individual enzyme molecule becomes soluble in the organic phase. Studies on the preparation of organic-soluble enzymes have proven that such a "homogeneous" distribution of polar enzyme molecules is possible. Moreover, in terms of activity, beneficial effects were observed, most likely explainable by less-restricted Brownian motion of the nanometer-sized complex in contrast to aggregates or immobilizates (16). Basically three different strategies to make organic-soluble enzymes can be distinguished: Enzymes can be modified (i) by the covalent binding of molecules like polyethylene glycol (PEG) or alkanoyl chlorides to enzyme surface, (ii) by the use of surfactants to encapsulate enzyme in the reversed micelles, or (iii) by the direct coating of enzyme molecules with a surfactant by ion-pairing. The PEG modification of enzymes is probably the best studied approach to enable homogeneous reactions with enzymes in organic media (17, 18). However, pegylated enzymes have usually limited solubility (typically below 1 mg/mL). Distel et al. (19) reported a novel method of modifying enzymes with short alkyl chains. A much better organic solubility (up to 44 mg/mL) has been achieved (19), though toxic reagent alkanoyl chloride was involved. Also the enzyme activity might get lowered by the covalent coupling as frequently reported in literature (20), the reason why this method does not work for all applications and enzymes.

Surfactant-assisted solubilization is the other direction to provide organic-soluble enzymes in the form of reverse micelles or as enzyme–surfactant complex formed through ion-pairing. Reversed micelles are usually formed under conditions that have surfactant concentration higher than the critical micelle concentration (CMC). But due to the relatively high surfactant concentration often the downstream processing is hampered (21). This particular problem is greatly minimized if the enzyme becomes soluble in organic media by direct ion-pairing with an oppositely charged surfactant, which is applied at a very low concentration (far below the CMC) compared to reverse micelles. Furthermore, unlike for reverse micelle almost anhydrous conditions can be achieved by ion-pairing (5). Technically the enzyme is solubilized in an aqueous phase and a small amount of surfactant in organic phase is added, followed by the extraction of the formed complex into an organic phase. A lot of research on the enzyme extraction from an aqueous into an organic phase using cationic and anionic surfactants was conducted (5, 8, 10, 11, 14, 21–27). Typically – depending on the enzyme and surfactant type – enzyme concentrations up to 200 mg/mL in the organic phase were obtained (unpublished data of Wang's Lab). Theoretically this intriguing technique can be applied to any enzyme either mono- or multimeric, but practically it is best suited to cheap enzymes. Because the recovery of a homogeneously solubilized enzyme is difficult, it normally has to be discarded after the completion of reaction.

Due to the electrostatic interaction between the oppositely charged functional groups of the enzyme molecule and the surfactant hydrophilic head group, an ion-paired enzyme–surfactant complex is formed. In the ideal case, with a size comparable to the enzyme molecule itself, the formed complex can be considered as nano-scaled (21). The hydrophobic tail of the surfactant is displayed on the surface of the complex, thus ensuring its solubility in the organic media and allowing for the homogeneous distribution of the whole complex (**Fig. 5.1**). Due to the gentle nature of electrostatic interactions involved, the enzyme's native structure remains unaffected; the activity/stability is typically retained as well (22). This chapter tends to guide the readers through the preparation of enzyme nano-complexes by ion-pairing with

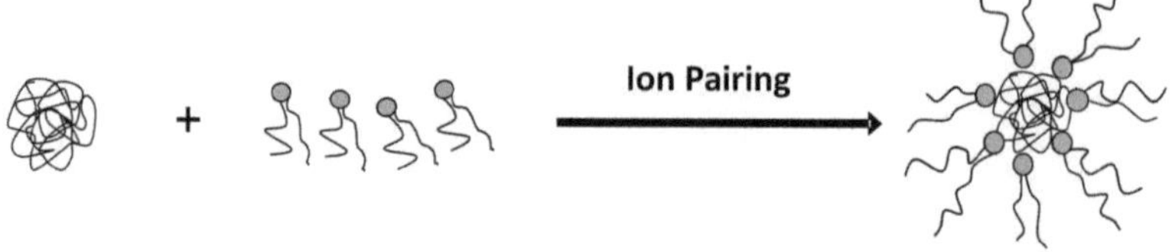

Fig. 5.1. Reaction scheme of ion-pairing to render enzymes organic soluble.

a surfactant. As model systems we selected two enzymes, α-chymotrypsin (α-CT) as a hydrolase and glucose oxidase (GO_x) as an oxidoreductase, to demonstrate the feasibility and generality of ion-pairing.

2. Materials

2.1. Solubilization of GO_x via Ion-Pairing

1. Surfactant solution for enzyme ion-pairing: Toluene containing 2 mM cationic surfactant dodecyl dimethylethyl ammonium bromide (DDAB) purchasable from common suppliers such as Sigma-Aldrich Co. (St. Louis, MO).
2. Sodium acetate buffer: 20 mM acetate, pH 5.5.
3. Enzyme (GO_x) solution: GO_x from *Aspergillus niger* (E.C. 1.1.3.4) purchased from Sigma-Aldrich Co. with relatively high purity (ca. 70%) is dissolved in sodium acetate buffer at a protein concentration of 0.7 mg/mL (*see* **Note 1**).
4. Protein standards: GO_x is dissolved in DI water at different concentrations for calibration (28).

2.2. Solubilization of α-CT via Ion-Pairing

1. Non-ionic bis–tris propane buffer: 10 mM, pH 7.8 containing 6 mM $CaCl_2$ (*see* **Note 2**).
2. Enzyme solution: Powder-like α-CT from bovine pancreas (E.C. 3.4.21.1) purchasable at Sigma-Aldrich Co. with high purity (ca. 85%) is dissolved in bis–tris propane buffer at a protein concentration of 1.1 mg/mL.
3. Extracting solution: Iso-octane containing 2 mM anionic surfactant docusate sodium salt (AOT) purchasable at Sigma-Aldrich Co.

2.3. Solutions for GO_x Activity Assay

1. Sodium acetate buffer: 50 mM; pH 5.1.
2. Chromogen solution: 10 mg of *o*-dianisidine dissolved in 1 mL DI H_2O.

3. Glucose solution for assay of GO_x activity: 10% (w/v) β-D(+)-glucose in DI H_2O.
4. Peroxidase solution: Peroxidase from horseradish purchasable at Sigma-Aldrich Co. is dissolved in DI H_2O at a concentration of 200 mg/mL for coupled reaction.

2.4. Solutions for α-Chymotrypsin Activity Assay

1. Substrate solution: Hexane containing *n*-acetyl-L-phenylalanine ethyl ester (APEE) purchasable at Sigma-Aldrich Co. with concentration ranging from 2.5 to 30 mM and 0.5 M of *n*-propyl alcohol (PrOH); hexane is pretreated with 3 Å zeolite molecular sieves for at least 24 h to remove undesired residue water.

3. Method

3.1. Solubilization of GO_x in Toluene via Ion-Pairing with a Cationic Surfactant

GO_x catalyzes the oxidation of β-D-glucose to δ-gluconolactone under the formation of H_2O_2. The resulting lactone is subsequently hydrolyzed, yielding gluconic acid. The most important applications of GO_x are in food preservation (biocidal effect of gluconic acid and H_2O_2) and biosensing (glucose determination in body fluids, beverages, and food stuff (29, 30)). GO_x is a homodimer of molecular weight 160 kDa in which each subunit contains a tightly bound FAD molecule as prosthetic group. The presence of FAD is also responsible for the yellow color of the enzyme (31) and allows for the easy spectrophotometric quantification of the enzyme at 450 nm. One great opportunity given by an organic-soluble GO_x is that the homogeneous embedment of the enzyme into a solid polymer becomes possible. Such a material can be denoted as bioplastic and allows the simplified assembly of, e.g., a biosensing device, while concomitantly enhancing the thermal stability of GO_x (15). The typical extraction procedure for the preparation of organic-soluble GO_x via ion-pairing is described as follows.

3.1.1. Choose Suitable pH of Aqueous Phase

Since electrostatic interactions play an important role in the extraction with ionic surfactants, the net charge of the enzyme must be considered first. Enzyme proteins are polyelectrolytes due to acidic and basic side chains of certain amino acids. As a function of pH these groups can be deprotonated or protonated, thus influencing the overall net charge of the enzyme protein. In this respect the isoelectric point (p*I*) is defined as the pH at which the protein has no net charge, which is normally accompanied by a lower solubility and stability. At a pH below p*I*, the basic side chains become protonated to give the protein a positive net charge, and vice versa. Therefore, for an efficient ion-pairing between the enzyme and the surfactant the pH of the aqueous phase should be chosen to ensure an oppositely charged enzyme and surfactant.

Considering the p*I* value of GO_x (4.2) and the optimum pH of native GO_x for the highest activity (pH 5.5 as evaluated by Wang's group (15)) it becomes obvious to use a cationic surfactant for extraction by ion-pairing. Therefore 0.02 M acetate buffer at pH 5.5 and DDAB as cationic surfactant are chosen for ion-pairing in this example (*see* **Note 3**).

3.1.2. Emulsification and Extraction Step

1. The aqueous GO_x solution (*see* **Section 2.1**) at the specified pH is poured into the toluene phase containing the cationic surfactant DDAB at a 1:1 phase ratio. Subsequently the mixture is vigorously stirred for 2 min at a constant rate of 300 rpm and a temperature of 25°C.
2. The emulsion (total volume of 20 mL) is transferred into 10× 2.0 mL microcentrifuge tubes for centrifugation. Centrifugation is done at 12,000 rpm for 3 min to ensure a complete phase separation. The successful extraction of GO_x becomes visible by the yellow color of the upper organic phase (*see* **Notes 4, 5,** and **6**). The upper organic phase is carefully removed from the centrifuge tubes.
3. The concentration of the extracted GO_x in the toluene phase can be determined spectrophotometrically at 450 nm based on the enzyme's prosthetic group FAD which gives the enzyme the specific yellow (*see* **Note 5**). The calibration is done by measuring the absorbance of aqueous solution with varying GO_x concentrations in DI water at the same wavelength. The extraction yield is determined as the ratio of total amount of extracted protein in the organic phase and the total amount of used protein.

3.1.3. Enrichment of the Enzyme Concentration in the Organic Phase

After the extraction step, the GO_x concentration is about 0.5 mg/mL. The enzyme concentration can be further increased by passing a stream of dry N_2 gas through the organic solvent when operated at a lab scale (*see* **Note 7**). It is even possible to remove the organic solvent completely – the organic-soluble enzyme remains as a powder than is easily resoluble in a wide range of organic solvents, either polar or non-polar (*see* **Note 8**), for various applications. Storage under refrigeration is recommended for the solid enzyme–surfactant complex.

The extraction efficiency depends on various factors like pH, the nature of the organic solvent, the enzyme type and concentration, as well as surfactant type and concentration. **Figure 5.2** depicts the impact of the pH of the aqueous phase and the surfactant concentration in the organic phase as the most obvious factors. As can be taken from **Fig. 5.2a**, extraction at a pH below

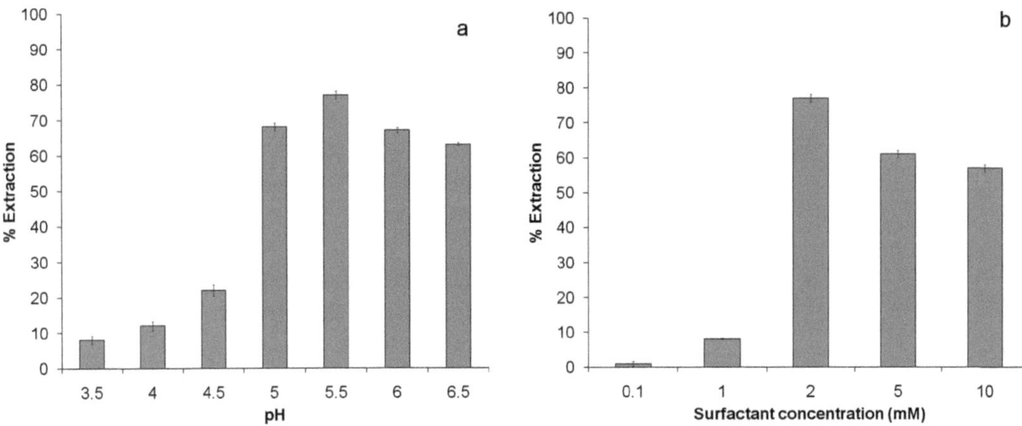

Fig. 5.2. Effects of pH and surfactant concentration on the extraction yield of GO_X. (**a**) Starting with 0.7 mg/mL GO_X in the aqueous phase, using toluene and 2 mM DDAB, higher extraction yields are obtained at pH values above the p*I* (4.2); the highest extraction yield was achieved at pH 5.5. (**b**) A surfactant concentration of 2 mM resulted in the highest extraction yield, while lower concentrations did not succeed extraction. It has to be noted that the used surfactant concentration is far below the CMC of DDAB (ca. 17 mM (32)).

p*I* of GO_x results in poor yields. The reason is that on the side of the enzyme there are barely negative charges available, which are needed for the attraction of the cationic surfactant. At pH 5.5 the maximum of GO_x is extracted. **Figure 5.2b** demonstrates the effect of DDAB on the extraction yield. Different DDAB concentrations 0.1–10 mM were studied, far below the CMC of DDAB in water–toluene (ca. 17 mM (32)). Based on our results the optimum concentration for the extraction of GO_x was determined to be 2 mM.

3.2. Solubilization of α-CT in Organic Solvent via Ion-Pairing

α-CT is a digestive enzyme that can perform proteolysis in aqueous reaction media. It has a molecular weight of 25 kDa. In the biocatalytic industry, α-CT is used for transesterifications and the synthesis of peptides, resulting in products that are, e.g., important as active pharmaceutical ingredients (23). In this respect the enzyme is typically used either as free powder or immobilizate. Organic-soluble α-CT will offer all of the aforementioned advantages, but especially as reported in literature a tremendously improved performance (5, 23).

The solubilization of α-CT by ion-pairing was first reported by Paradkar and Dordick (5) and further investigated by other researchers (22, 23). The basic procedure for the extraction of α-CT into an organic phase is very similar to what was described for the extraction of GO_X. However, due the much higher p*I* of α-CT (8.8) compared to GO_x (4.2), here the ion-pairing required the use of an anionic surfactant like AOT. Bis–tris propane buffer at a pH of 7.8 is used. Under these conditions the enzyme is positively charged and able to attract the negatively charged surfactant.

3.2.1. Emulsification and Extraction Step

1. The aqueous α-CT solution is poured into toluene containing 2 mM AOT and vigorously shaken for 2 min at a constant stirring rate of 250 rpm at 25°C.
2. The emulsion (total volume of 20 mL) is transferred into 10× 2.0 mL microcentrifuge tubes for centrifugation. Centrifugation is done at 12,000 rpm for 3 min to ensure a complete phase separation. The upper organic phase is carefully removed from the centrifuge tubes. The concentration of α-CT in the organic phase is determined spectrophotometrically at 280 nm (*see* **Note 3**).

3.2.2. Enrichment of the Enzyme Concentration in the Organic Phase

After extraction, the protein concentration can be increased by passing a stream of dry N_2 gas through the organic solvent. In this case the organic-soluble α-CT remains as a transparent sticky matter that can be easily resolubilized in the organic solvents, but also here a lower activity can be expected with polar organic solvent, which can be explained by the stripping of essential hydrate shell of the organic-soluble enzyme as observed in enzymes suspended as powder in similar solvents (5) (*see* **Note 8**). Storage in the refrigerator is recommended for the concentrated complex. As reported by Paradkar and Dordick (5), the organic-soluble α-CT–surfactant complex retains its native structure in the organic solvents, suggesting that the ion-pairing modification is a gentle method with no significant loss of activity. In contrast to the lyophilized counterpart which is just suspended in the organic solvent, these authors reported an activity for the organic-soluble enzymes which was 2–3 orders of magnitude greater.

3.3. Characterization of the Enzyme–Surfactant Complex

Once formed the enzyme–surfactant complex can be characterized by a variety of methods. The most interesting parameters to know are the size of the complex, the ratio of surfactant/extracted protein, the water content, and how they are affecting the enzyme's structure and activity. The size of dehydrated enzyme–surfactant complex in iso-octane can be determined by dynamic light scattering. Paradkar and Dordick (5) reported a mean size of 6.8 nm for the α-CT–AOT complex, which is roughly the same as for the unmodified native α-CT.

Extracted protein concentration could be determined by UV spectrometry as discussed previously. The mass ratio of surfactant per extracted protein can be roughly estimated if the mass balance of the final weight of the dried organic-soluble complex and the amount of the extracted enzyme is made. For instance, it was estimated that the number of surfactant molecules per extracted α-CT molecule adds up to ca. 30, which is far away from the number of surfactant molecules necessitated for forming a reverse micelle (5).

The water content of the final organic-soluble complex might be of importance for certain biocatalytic reactions like, for example, esterifications and can be determined by Karl Fischer titration (10). Typically the water content in the final complex is lower than 0.1% v/v. Possible changes of the enzyme confirmation upon ion-pairing can be monitored by circular dichroism (CD) (21) and fluorescent spectroscopy (5). As was shown in previous studies no major changes are to be expected.

3.3.1. Specific Activities of Free and Solubilized GO_x

One widely used spectrophotometric method to determine the catalytic activity of glucose oxidase is a peroxidase-coupled enzyme reaction, based on the oxidation of *o*-dianisidine through H_2O_2 which is formed as side product during the oxidation of glucose (33). This assay is well suited to judge the performance in respect of activity and stability of the enzyme–surfactant complex. For a comparative base the measured catalytic activity is related to the amount of enzyme used, denoted as specific activity (U per mg protein). Due to the presence of FAD in the enzyme molecule, it will show an absorbance maximum at 450 nm, which can be used to quantify the enzyme amount either in the aqueous or in the organic phase.

The activity assay is started by the dispersion of 10 μL of the organic-soluble GO_x (solubilized in toluene at suitable concentrations) in a mixture of the following solutions: 2.25 mL sodium acetate buffer (50 mM; pH 5.1), 17 μL *o*-dianisidine solution, 534 μL 10% (w/v) β-D(+)-glucose solution, and 105 μL peroxidase solution. It is recommended to conduct the reaction in a 4 mL disposable cuvette since it allows the on-line monitoring of the reaction in the spectrophotometer. The activity could be measured by recording the product absorbance reading at 500 nm continuously within 5 min. It is recommended to conduct the experiment at least in triplicate to obtain reliable results.

3.3.2. Specific Transesterification Activities of Free and Solubilized α-CT

The enzyme α-CT catalyzes the transesterification of APEE with 1-proponaol in organic solvents. This reaction is used to determine the catalytic activity of organic-soluble α-CT compared to their unmodified counterpart. The reaction is carried out at room temperature and started by adding specified amounts of either the modified or the native enzyme to 10 mL hexane reaction solution containing APEE and 1-propanol as well as a small amount of water (<1%). The native α-CT is suspended as powder, while the organic-soluble α-CT is homogeneously solubilized. After the enzyme is added the reaction mixture is vigorously shaken at 200 rpm. Samples of 200 μL for GC analysis are withdrawn periodically from the organic phase in course of the reaction. While the native powder-like enzyme can be removed easily by filtration or centrifugation to stop the reaction, the organic-soluble enzyme needs to be inactivated. According to literature this can be done by the addition of 20 μL glacial acetic acid. The product formation can be monitored by using a gas chromatograph equipped with a FID detector and a RTX-5 capillary column (0.25 mm × 0.25 μm × 10 m, Shimadzu). A temperature gradient from 100 to 190°C at a heating rate of 20°C/min, followed by 5-min holding at 190°C, is recommended as well as an injector and detector temperature of 210 and 280°C, respectively. Obtained data are used to determine the initial reaction rate of the formation of *n*-acetyl-L-phenylalanine propyl ester (APPE).

4. Notes

1. It cannot be excluded that the efficiency of the extraction process is impaired by impurities contained in the enzyme purification. For that reason we suggest to start the ion-pairing with an enzyme preparation of the highest purity available. If necessary, a purification step such as by ultrafiltration is recommended.
2. For certain applications it might be beneficial to use a non-ionic buffer (e.g., bis–tris propane buffer). Some common buffers, especially those made of relatively large charged molecules, might interfere with the extraction process. When a non-ionic buffer is used, the ionic strength can be controlled by the addition of salts like $CaCl_2$.
3. There is no thumb of rule for the selection of the right surfactant for a certain enzyme or application. Once an enzyme is chosen for ion-pairing, generally a thorough literature search is recommended to exclude at least those surfactants that have been proven unfavorable. **Table 5.1** lists different surfactants for various enzymes that have been used for ion-pairing. For the choice of the cationic or anionic surfactant the net charge of the enzyme at a certain pH must be considered. According to literature predominantly anionic surfactants have been used for ion-pairing so far; AOT can be the first choice to start. A variety of protocols are available using anionic surfactant AOT to complex various types of enzymes. Although cationic surfactants are also applicable, generally they turned out to be less advantageous. One reason is given by the fact that many enzymes have a positive net charge at the optimum pH (21).
4. The spectrometry analysis at 280 nm is used to determine the organic-soluble protein concentration in the organic

Table 5.1
Extraction yields for various enzymes using different surfactants and solvents

Enzymes (p*I*)	pH	Surfactant and solvent	Extraction yield (%)	References
Bacillus subtilis α-amylase (7.0)	6.5	AOT in hexane	6	Unpublished data (Wang's Lab)
Aspergillus niger glucose oxidase (4.2)	5.5	DDAB in toluene	70	Unpublished data (Wang's Lab)
Bacillus subtilis protease N (7.0)	6.5	AOT in hexane	30	Unpublished data (Wang's Lab)
Subtilisin Carlsberg (9.4)	7.5	AOT in hexane	20	Unpublished data (Wang's Lab)
Lipase B (N/A)	(N/A)	Didodecyl glucosylglutamate in iso-octane	50	(25)
Mucor javanicus lipase (6.0)	5.0	AOT in iso-octane	35	(8)
Candida rugosa lipase (5.0)	4.5	AOT in iso-octane	32	(26)
Horseradish peroxidase (7.2)	6.6	AOT in iso-octane	65	(27)
Horseradish peroxidase (5.2)	3.0	AOT in iso-octane	94	(11)
Bovine pancreas α-chymotrypsin (8.8)	7.8	AOT in iso-octane	90	(5)

solvent (34), provided that the solvents and surfactants used do not interfere. A proof scan is recommended for solvents and surfactants to ensure that there is no interference. The calibration curve can be done with a standard of any pure protein like bovine serum albumin (BSA).

5. If an aromatic solvent like toluene is used for the extraction, the above-mentioned spectrophotometric protein quantification at 280 nm cannot be applied due to interference at the specified wavelength. If a protein or enzyme with a chromophore is used, it can be used for monitoring the extracted protein amounts in the upper phase. As in the present case GO_x can be quantified at 450 nm by monitoring the specific yellow color of FAD that is attached to the enzyme. If aromatic solvents and enzymes without a chromophore are used, the extraction process can be monitored by a decrease of the protein concentration in the aqueous phase via the Bradford assay (35).
6. For a correct quantification of the extracted protein amounts, it is necessary to determine the precise volume of the upper organic phase. Therefore the upper phase has to

be removed carefully after centrifugation without touching the interface or the lower aqueous phase. Alternatively the volume/weight of the added organic phase can be monitored carefully. Of course, the correct quantification makes it necessary to avoid the evaporation of low-boiling organic solvents in course of the extraction process.

7. For up-scaling of the ion-pairing method, the recycling of the bulk organic solvent by a rotary evaporator is strongly recommended.

8. Depending on the nature of the surfactants and enzymes, as well as preparation conditions, the final complexes show various forms after drying. Typically, the dried GO_x–DDAB complex is a yellow powder, while the dried α-CT–AOT complex appears as a transparent plastic-like solid. Both are, in respect of activity, very stable at ambient conditions. Resolubilization upon the addition of non-polar organic solvents caused no significant activity loss, although a detectable activity loss is registered during resolubilization in some polar solvents (e.g., methanol).

References

1. Klibanov, A. M. (2001) Improving enzymes by using them in organic solvents. *Nature* **409**, 241–246.
2. Zaks, A., and Klibanov, A. M. (1985) Enzyme-catalyzed processes in organic solvents. *PNAS* **82**, 3192–3196.
3. Zaks, A., and Klibanov, A. M. (1988) Enzymatic catalysis in nonaqueous solvents. *J. Biol. Chem.* **263**, 3194–3201.
4. Zhao, X., El-Zahab, B., Brosnahan, R., Perry, J., and Wang, P. (2007) An organic soluble lipase for water-free synthesis of biodiesel. *Appl. Biochem. Biotechnol.* **143**, 236–243.
5. Paradkar, V. M., and Dordick, J. S. (1994) Aqueous-like activity of alpha-chymotrypsin dissolved in nearly anhydrous organic solvents. *J. Am. Chem. Soc.* **116**, 5009–5010.
6. Ferjancic, A., Puigserver, A., and Gaertner, H. (1990) Subtilisin-catalysed peptide synthesis and transesterification in organic solvents. *Appl. Microbiol. Biotechnol.* **32**, 651–657.
7. Wu, W. H., Akoh, C. C., and Phillips, R. S. (1996) Lipase-catalyzed stereoselective esterification of menthol in organic solvents using acid anhydrides as acylating agents. *Enzyme Microb. Technol.* **18**, 536–539.
8. Altreuter, D. H., Dordick, J. S., and Clark, D. S. (2002) Optimization of ion-paired lipase for non-aqueous media: Acylation of doxorubicin based on surface models of fatty acid esterification. *Enzyme Microb. Technol.* **31**, 10–19.
9. Ikeda, I., and Klibanov, A. M. (1993) Lipase-catalyzed acylation of sugars solubilized in hydrophobic solvents by complexation. *Biotechnol. Bioeng.* **42**, 675–791.
10. Bruno, F. F., Akkara, J. A., Ayyagari, M., Kaplan, D. L., Gross, R., Swift, G., and Dordick, J. S. (1995) Enzymic modification of insoluble amylose in organic solvents. *Macromolecules* **28**, 8881–8883.
11. Angerer, P. S., Studer, A., Witholt, B., and Li, Z. (2005) Oxidative polymerization of a substituted phenol with ion-paired horseradish peroxidase in an organic solvent. *Macromolecules* **38**, 6248–6250.
12. Yang, F., and Russell, A. J. (1995) A comparison of lipase-catalyzed ester hydrolysis in reverse micelles, organic solvents, and biphasic systems. *Biotechnol. Bioeng.* **47**, 60–70.
13. Wang, P., Sergeeva, M. V., Lim, L., and Dordick, J. S. (1997) Biocatalytic plastics as active and stable materials for biotransformations. *Nat. Biotechnol.* **15**, 789–793.
14. Novicka, S. J., and Dordick, J. S. (2002) Protein-containing hydrophobic coatings and films. *Biomaterials* **23**, 441–448.
15. Wang, P., Narayanan, R., Wu, S., Jia, H., and Renker, D. H. (2009) Nanofibers with high enzyme loading for highly

sensitive biosensors. US Patent Application: US2008009835 (Publication date: May 3, 2009).

16. Jia, H., Zhu, G., and Wang, P. (2003) Catalytic behaviors of enzymes attached to nanoparticles: The effect of particle mobility. *Biotechnol. Bioeng.* **84**, 406–414.
17. Ohya, Y., Sugitou, T., and Ouchi, T. (1995) Polycondensation of a-hydroxy acids by enzymes or peg-modified enzymes in organic media. *Pure Appl. Chem. A* **32**, 179–190.
18. Wang, P., Woodward, C. A., and Kaufman, E. N. (1999) Poly(ethylene glycol)-modified ligninase enhances pentachlorophenol biodegradation in water-solvent mixtures. *Biotechnol. Bioeng.* **64**, 290–297.
19. Distel, K. A., Zhu, G., and Wang, P. (2005) Biocatalysis using an organic-soluble enzyme for the preparation of poly(lactic acid) in organic solvents. *Bioresource Technol.* **96**, 617–623.
20. Martinek, K., Klibanov, A. M., Goldmacher, V. S., and Berezin, I. V. (1977) The principles of enzyme stabilization. I. Increase in thermostability of enzymes covalently bound to a complementary surface of a polymer support in a multipoint fashion. *Biochim. Biophys. Acta* **485**, 1–12.
21. Meyer, J. D., and Manning, M. C. (1998) Hydrophobic ion pairing: Altering the solubility properties of biomolecules. *Pharm. Res.* **15**, 188–193.
22. Meyer, J. D., Matsuura, J. E., Kendrick, B. S., Evans, E. S., Evans, G. J., and Manning, M. C. (1995) Solution behavior of alpha-chymotrypsin dissolved in nonpolar organic solvents via hydrophobic ion pairing. *Biopolymers* **35**, 451–456.
23. Altreuter, D. H., Dordick, J. S., and Clark, D. S. (2003) Solid-phase peptide synthesis by ion-paired alpha-chymotrypsin in nonaqueous media. *Biotechnol. Bioeng.* **81**, 809–817.
24. Kim, J., Kosto, T. J., II, Manimala, C. J., Nauman, B. E., and Dordick, J. S. (2001) Preparation of enzyme-in-polymer composites with high activity and stability. *AICHE J.* **47**, 240–244.
25. Tsuzuki, W., Okahata, Y., Katayama, O., and Suzuki, T. (1991) Preparation of organic-solvent-soluble enzyme (lipase B) and characterization by gel permeation chromatography. *J. Chem. Soc., Perkin Trans.* **1**, 1245–1275.
26. Wu, J. C., Chua, P. S., Chow, Y., Li, R., Carpenter, K., and Yu, H. (2006) Extraction of *Candida rugosa* lipase from aqueous solutions into organic solvents by forming an ion-paired complex with AOT. *J. Chem. Technol. Biotechnol.* **81**, 1003–1008.
27. Bindhu, L. V., and Emilia Abraham, T. (2003) Preparation and kinetic studies of surfactant–horseradish peroxidase ion paired complex in organic media. *Biochem. Eng. J.* **15**, 47–57.
28. Gouda, M. D., Singh, S. A., Rao, A., Thakur, M. S., and Karanth, N. G. (2003) Thermal inactivation of glucose oxidase: Mechanism and stabilization using additives. *J. Biol. Chem.* **278**, 24324–24333.
29. Hecht, H. J., Kalisz, H. M., Hendle, J., Schmid, R. D., and Schomburg, D. (1993) Crystal structure of glucose oxidase from Aspergillus niger refined at 2·3 Å resolution. *J. Mol. Biol.* **229**, 153–172.
30. Turner, A., Karube, I., and Wilson, G. S. (1987) *Biosensors: Fundamentals and Applications.* Oxford University Press, Oxford, p. 770.
31. Haouz, A., Twist, C., Zentz, C., Tauc, P., and Alpert, B. (1998) Dynamic and structural properties of glucose oxidase enzyme. *Eur. Biophys. J.* **27**, 19–25.
32. Mehta, S. K., Bhasina, K. K., Chauhana, R., and Dhama, S. (2005) Effect of temperature on critical micelle concentration and thermodynamic behavior of dodecyldimethylethylammonium bromide and dodecyltrimethylammonium chloride in aqueous media. *Colloid Surf. A* **255**, 153–157
33. Wei, Y., Dong, H., Xu, J., and Feng, Q. (2002) Simultaneous immobilization of horseradish peroxidase and glucose oxidase in mesoporous sol-gel host materials. *Chem Phys. Chem.* **3**, 802–808.
34. Stoscheck, C. M. (1990) Quantitation of protein. *Methods Enzymol.* **182**, 52–69.
35. Bradford, M. M. (1976) A rapid and sensitive method for the quantitation of microgram quantities of protein utilizing the principle of protein-dye binding. *Anal. Biochem.* **72**, 248–254.

Chapter 6

Enzyme-Immobilized CNT Network Probe for In Vivo Neurotransmitter Detection

Gi-Ja Lee, Seok Keun Choi, Samjin Choi, Ji Hye Park, and Hun-Kuk Park

Abstract

Glutamate is the principal excitatory neurotransmitter in the brain, and its excessive release plays a key role in neuronal death associated with a wide range of neural disorders. Real-time monitoring of extracellular glutamate levels would be very helpful in understanding the excitotoxic process of neurotransmitters on brain injury. Toward the detection of L-glutamate, we describe in this chapter the preparation of carbon nanotube (CNT) network probes with immobilized L-glutamate oxidase (GLOD) by using a non-covalent functionalized method. Such GOLD-CNT network probes are evaluated with real-time electronic responses corresponding to standard glutamate solutions in vitro and a 11-vessel occlusion (11 VO) rat model in vivo. The ultrahigh sensitivity, selectivity, and fast response time of GLOD-CNT network probes are greatly promising for the real-time electronic detection of extracellular glutamate levels in brain.

Key words: Glutamate, CNT network probe, in vivo, real-time detection, GLOD immobilization.

1. Introduction

A field-effect transistor (FET) is a family of transistor commonly used to amplify electronic signals. In the FET, current flows along a conducting channel. Two electrodes at both ends of the channel are called the source and the drain, respectively. Electrical conductivity of the channel can be varied by the application of a voltage to a control electrode called the gate. A small change in gate voltage can cause a large variation in the current from the source to the drain. FETs which are fabricated using semiconducting single-walled carbon nanotube (SWNT) as the conducting channel are

P. Wang (ed.), *Nanoscale Biocatalysis*, Methods in Molecular Biology 743,
DOI 10.1007/978-1-61779-132-1_6, © Springer Science+Business Media, LLC 2011

called nanotube FET (1). Since nanotube FETs were measured first time in 1998 (2, 3), great progress has been made in developing nanotube-based FET sensors for chemical and biological applications (4–7). FET made of carbon nanotube (CNT) can be used as an excellent sensor due to its extreme sensitivity to local chemical environments (8). The high sensitivity of CNT originates from its atoms that are located at the surface. Proteins such as enzymes and antibodies which are specific to each target have been used as recognition elements. It is essential to immobilize biological molecules on CNTs by a reliable method to preserve their electronic characteristics. The functionalization of CNT is a prerequisite for the immobilization of biomolecules. CNT functionalization is divided into covalent and noncovalent methods. Covalent functionalization of CNTs is accomplished by three different approaches – thermally activated chemistry, electrochemical modification, and photochemical functionalization (9). As covalent functionalization makes it difficult to keep the sp^2 structure and property of CNT, it was preferred to utilize the noncovalent sidewall functionalization of CNT. Chen et al. (10) reported that the sidewalls of single-walled carbon nanotube (SWNT) were noncovalently functionalized using bifunctional molecule such as pyrenyl group and Shim et al. (11) co-adsorbed the Triton surfactant and polymer species on CNTs for the sidewall modification. In 2003, Besteman et al. (12) found that immobilization of glucose oxidase (GOD) on the sidewall of a semiconducting single-walled carbon nanotube (SWNT) decreased the conductance of the tube. As a result, they suggested that an enzyme activity sensor could be constructed at the single molecule level of an individual SWNT.

Excitatory amino acids have been proposed to play a major role in the development of neuronal damage during and following cerebral ischemia (13). It is reported that the level of extracellular glutamate is closely related to the reduction of cerebral blood flow (CBF) (14) and brain temperature (15) during ischemia. It has been demonstrated that massive glutamate elevation occurs in severe global ischemia (16) and at the core in focal ischemia (17), while with mild or moderate global ischemia (18) and in the ischemic penumbra with focal ischemia (19). Therefore, a better understanding of excitotoxic process is required for an accurate assessment of changes in extracellular glutamate levels during and after the various insults suspected of initiating this type of damage.

In previous studies, the extracellular concentrations of glutamate and K^+ in the brain have been measured by high-performance liquid chromatography coupled with the microdialysis method and a K^+-sensitive electrode (20, 21). However, since it takes a long time to collect a sufficient volume of sample for the test, the real-time detection of the analytes in the early period of ischemia is not possible. And, because traditional microdialysis methods are limited by the time constraints of sampling and dilution of the analytes, it can grossly underestimate the concentration of excitatory amino acids in the extracellular space. Albery et al. (22) developed a dialysis electrode system which consisted of electrode and a standard microdialysis probe containing glutamate oxidase (GLOD). In this method, the glutamate is oxidized by GLOD in perfusate and produced hydrogen peroxide which can be amperometrically detected in real time on a platinum working electrode in the tip of the probe. This dialysis electrode has been used in a study of the rat cerebral ischemia model. But the dialysis electrode with low sensitivity and moderate response time makes it difficult to detect the rapid and minute changes in the glutamate level within the extracellular space. For real-time monitoring of extracellular glutamate levels, development of new sensor with high sensitivity, selectivity, and fast response time is required.

In this chapter, we report the methods of real-time glutamate detection by GLOD-coated CNT network probe in the in vitro and in vivo (11-vessel occlusion rat model) tests.

2. Materials

1. The probe-type CNT network junctions: supply from the Nano Lab., School of Physics and Astronomy, Seoul National University, by surface-programmed assembly method (23, 24).
2. Modification of CNT network junctions for enzyme immobilization: a 1 mM 1-pyrenebutanoic acid succinimidyl ester solution in methanol.
3. A 10 mM phosphate buffer saline (PBS) solution which consists of sodium phosphate, monobasic and dibasic; sodium chloride; and potassium chloride: pH 7.4, store at 4°C.
4. Enzyme immobilization on CNT network junctions: a 0.1 U/μL L-glutamate oxidase (50 U/vial, Yamasa Corp., Tokyo, Japan) solution in 10 mM PBS. Store at 4°C (*see* **Note 1**)
5. Deactivation of the excess reactive groups and prevention of nonspecific binding on CNT network junctions: a 100 mM ethanolamine solution in 10 mM PBS.
6. A L-glutamic acid standard solution in 10 mM PBS.
7. The measurement of electronic response from CNT network probe: DC power supply and picoammeter.

8. Animal model for in vivo detection: male Sprague-Dawley (SD) rats (250–300 g).
9. Real-time monitoring system for in vivo test: laser Doppler flowmetry for CBF measurement, 511 AC amplifier for electroencephalography (EEG), homeostatic blanket control unit for the maintenance of body temperature, DASY-Lab (Data Acquisition System, National Instruments, TX, USA) for various signal acquisition, and DADiSP 2002 (DSP Development Co., MA, USA) for data analysis.

3. Methods

The CNT network probe is based on the change in current between the source and the drain by its chemical environments. For the detection of glutamate, it is necessary to immobilize GLOD on CNT. First, bifunctional molecule, 1-pyrenebutanoic acid succinimidyl ester, adsorbs onto the surface of CNTs. The pyrenyl group adsorbs onto the hydrophobic surfaces of CNTs via $\pi-\pi$ interaction and the succinimidyl ester group reacts with amines which exist on the surface of GLOD, like other proteins (10). After GLOD immobilization, ethanolamine deactivates and blocks the remaining reactive groups for prevention of nonspecific binding onto CNT surface. Hydrogen peroxide and ammonia molecule are produced in the catalytic reaction between glutamate and GLOD. The products by glutamate cause the changes in conductance of CNT.

3.1. Functionalization of CNTs and Enzyme Immobilization on CNT

1. The CNT probes are incubated for 18 h with 1 mM 1-pyrenebutanoic acid succinimidyl ester in methanol at room temperature while stirring at 300 rpm, then wash with methanol for 2 h while stirring at the same speed (10) (*see* **Note 2**).
2. The probes are exposed to the GLOD solutions in 10 mM PBS (pH 7.4) at 4°C for 18 h.
3. After rinsing with clean PBS, in order to deactivate and block the excess reactive groups remaining on the surface, the 100 mM ethanolamine is added onto the probes and incubated for 30 min.
4. After re-rinsing with clean PBS, store GLOD-immobilized CNT probe at 4°C in refrigerator before usage (*see* **Note 3**).

3.2. In Vitro Measurements

1. After washing with PBS, the GLOD-immobilized CNT probe is dipped in 20 mL of fresh buffer solution for electrical measurements at room temperature.

2. A drain-source voltage (V_{sd}) of 10 mV is applied by DC power supply and the conductance change is measured with picoammeter. The experimental setup for detection of glutamate using CNT network probe in aqueous solution is shown in **Fig. 6.1**.
3. The liquid gate voltage is applied to 0 V using platinum wire electrode.
4. Changes of current between source and drain electrodes (I_{sd}) are measured when L-glutamic acid is subsequently added to the probe with or without GLOD. Examples of the results produced are shown in **Figs. 6.2** and **6.3** (*see* **Note 4**).

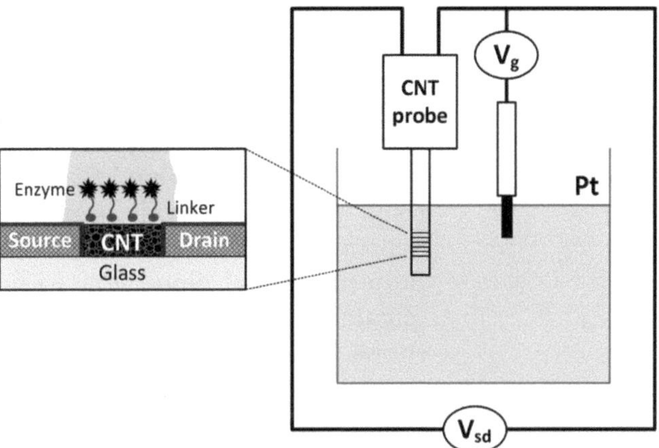

Fig. 6.1. Schematic diagram of the experimental setup for sensing of CNT network probe in aqueous solution (in vitro).

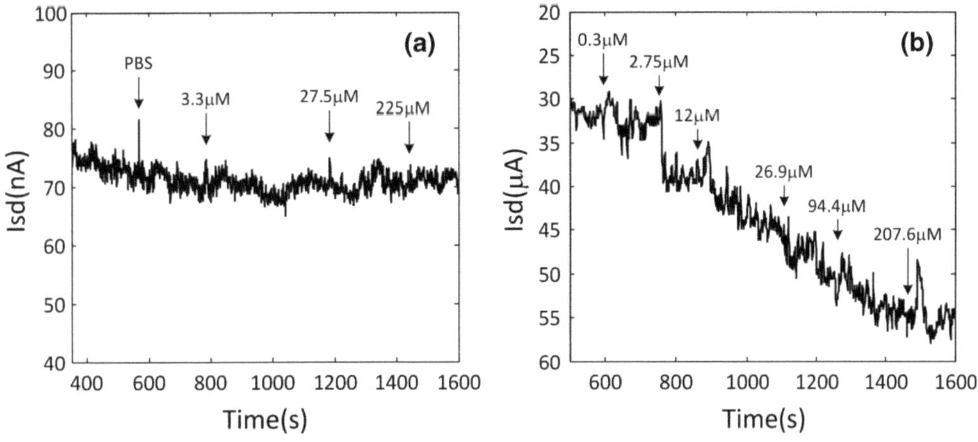

Fig. 6.2. (**a**) Real-time electronic response of bare CNT network transistor without GLOD in 20 mL PBS (pH 7.4) solution (V_g=0 V). (**b**) The conductance changes of the CNT network transistor with GLOD when the L-glutamate was dropped in 20 mL PBS (pH 7.4) solution (V_g=0 V).

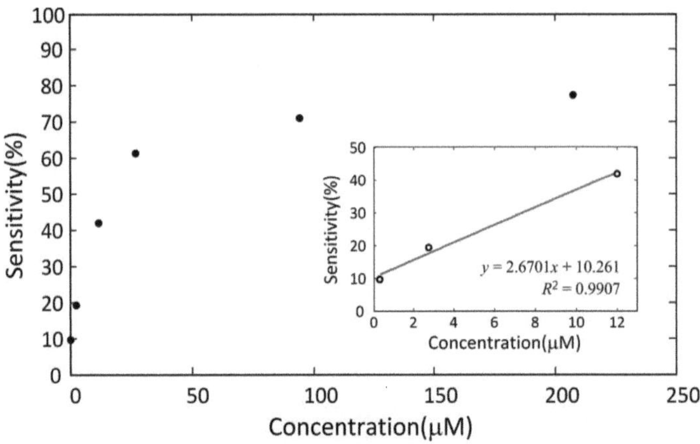

Fig. 6.3. The sensitivity of CNT network transistor as a function of glutamate concentration. The *inset* shows the linear response regime within the 300–15,000 nM concentration range.

3.3. Animal Stroke Model for In Vivo Detection

For real-time in vivo glutamate measurements, the 11-vessel occlusion (VO) rat models are utilized (25). Ischemia model using 11 VO methods is well established as a severe global ischemia (26). The 11 VO model has advantages of immediate drop in cerebral blood flow to near-zero levels and maximized glutamate release maintaining self-respiration (25, 27), which shows mimicking physiological forebrain ischemia compared to the traditional 4 VO technique. In 11 VO cerebral ischemia model, the CBF decreases rapidly to near 10% of the original levels as soon as the onset of ischemia, in coincident with the development of a flat electroencephalography signal. These changes are maintained throughout the 10-min ischemic period. When reperfusion is initiated, cerebral blood flow significantly increases to pre-ischemic levels (27).

1. SD rats are anesthetized with chloral hydrate (0.1 cc/100 mg with intraperitoneal injection).
2. 11 VO rats are prepared by division of the omohyoid muscle to allow the larynx to be retracted, exposing the ventral surface of the clivus following retraction of longus coli muscle.
3. 3 mm craniotomy is drilled through the clivus, centered just caudal to the basioccipital suture.
4. After opening the dura and arachnoid, basilar artery is coagulated just caudal to the origin of anterior inferior cerebellar artery.
5. Pterygopalatine arteries are coagulated prior to the entrance into the tympanic bullae.
6. Occipital arteries, superior thyroid arteries, and ascending pharyngeal arteries are identified, coagulated, and transected.

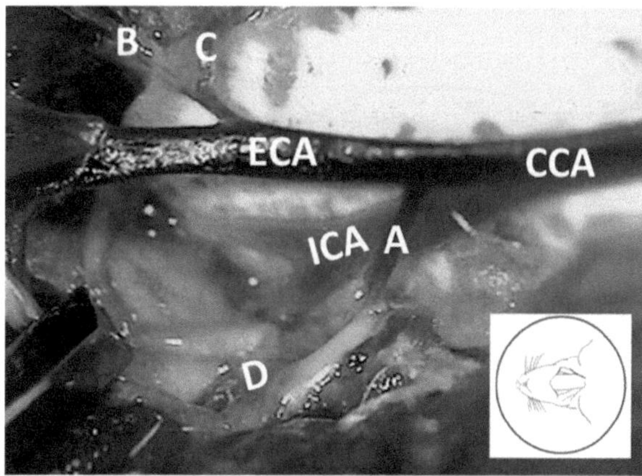

Fig. 6.4. Vascular anatomy of the neck dissection at right side. The adjacent musculature was removed. Occipital artery (*A*) is originated from external carotid artery (*ECA*). Ascending pharyngeal artery (*B*) and superior thyroid artery (*C*) are branched with common trunk. Pterygopalatine artery (*D*) is raised from internal carotid artery (*ICA*). *CCA*: common carotid artery.

7. Snares are placed around external carotid arteries, between occipital arteries proximally and superior thyroid arteries distally, and are placed on common carotid arteries. The vascular anatomy of the neck dissection is shown in **Fig. 6.4**.

3.4. Real-Time Monitoring of Glutamate Release in Stroke Model

1. Animals are placed in stereotaxic head holder for real-time glutamate monitoring.
2. On prone position in the stereotaxic frame, the skin of specimen is incised with midline along the superior sagittal suture.
3. Eight burr holes are made; at 4 mm apart from the midline and 1 mm front of coronal suture, one for the glutamate probe insertion and another for brain temperature measurement.
4. At 4 mm apart from the sagittal suture and 1 mm behind of coronal suture, two burr holes are made for EEG electrode at both sides.
5. At the 2 mm behind point of EEG site, two burr holes are made for CBF probe.
6. Finally two burr holes for EEG are made 2 mm behind of CBF probe at both sides.
7. Changes in CBF are monitored by laser Doppler flowmetry.
8. Body temperature is maintained at 36.5 ± 0.5°C with Homeostatic Blanket Control Unit.
9. Brain temperature is maintained at 37.5 ± 1°C through epidural thermocouple connected PID controller.

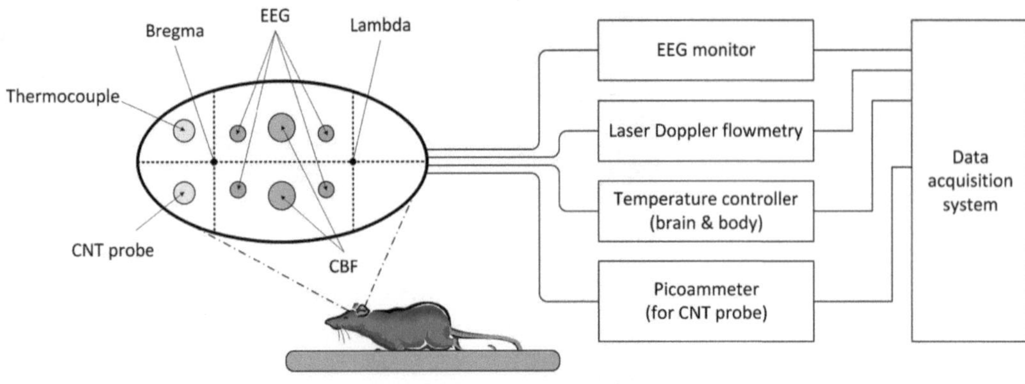

Fig. 6.5. The diagram of experimental setup for real-time glutamate monitoring system in 11-vessel occlusion rat model. *Zoomed image* shows the coordinates for craniotomy.

10. The CNT probes with and without GLOD are inserted into the motor cortex at coordinates A 1; L 4 (from the bregma); V 2 mm (from the dura) (*see* **Note 5**).
11. Data acquisition system is worked on, as shown in **Fig. 6.5**.
12. After a few minutes stabilization, a 10 min 11 VO cerebral ischemia is initiated by pulling the snares on CCAs and ECAs.
13. The snares are released and withdrawn after 10 min.
14. Examples of real-time monitoring of cerebral glutamate levels and CBF levels in the 11 VO rat model are shown in **Figs. 6.6** and **6.7** by CNT probes without GLOD and with GLOD, respectively (*see* **Notes 6** and 7).

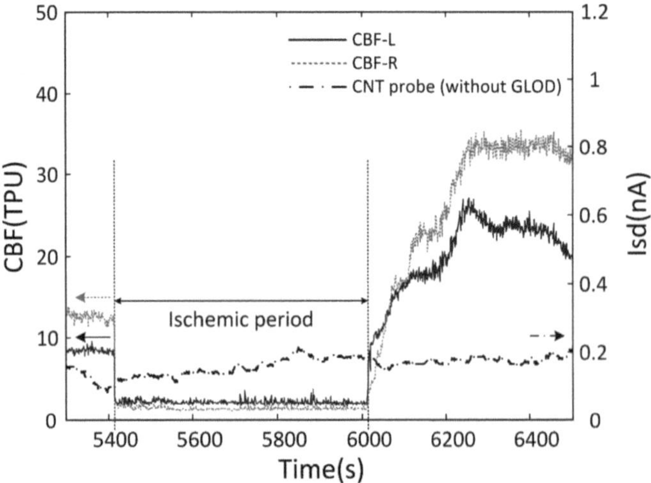

Fig. 6.6. Real-time changes of extracellular glutamate levels and CBF levels in 11 VO rat model by CNT network probe without GLOD.

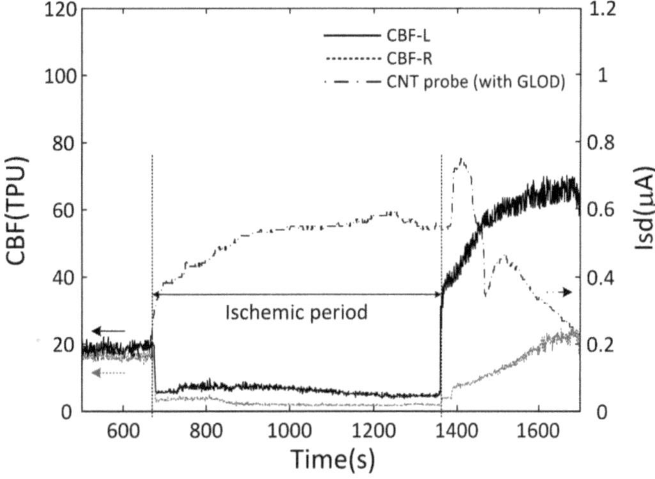

Fig. 6.7. Real-time changes of extracellular glutamate levels and CBF levels in 11 VO rat model by CNT network probe with GLOD.

4. Notes

1. GLOD solution in PBS must be stored at 4°C in refrigerator and used within 1 week.
2. Close the beaker when the CNT probes are incubated overnight with 1 mM 1-pyrenebutanoic acid succinimidyl ester in methanol at room temperature while stirring.
3. The enzyme-immobilized CNT probe must be stored at 4°C in refrigerator and used within 1 week.
4. From the changes in the I_{sd} of CNT probe before and after the immobilization of GLOD, we confirm that only the CNT probe with the activity of GLOD shows the specific response to glutamate.
5. After insertion in brain of 11 VO rat model, the CNT network probe is fully stabilized until background current is unchanged.
6. After the onset of ischemia in 11 VO model, the conductance of the CNT probe immediately increases. And it continues to increase throughout the entire ischemic period. When reperfusion initiates, it decreases to the pre-ischemic levels during reperfusion.
7. The CNT probes have great possibilities for in vivo neurotransmitter detection due to their rapid response and high sensitivity.

Acknowledgments

This study was supported by the research fund from the Ministry of Commerce, Industry, and Energy (MOCIE Grant #10017190-2007-31), and Bio R&D program through MEST (2010-0019912). We greatly appreciate Mr. B.Y. Lee, Mr. D.H. Son, and Prof. S. Hong at Seoul National University who provided the CNT probes.

References

1. Gruner, G. (2006) Carbon nanotube transistors for biosensing applications. *Anal. Bioanal. Chem.* **384**, 322–335.
2. Tans, S. J., Verschueren, A. R. M., and Dekker, C. (1998) Room-temperature transistor based on a single carbon nanotube. *Nature (London)* **393**, 49–52.
3. Martel, R., Schmidt, T., Shea, H. R., Hertel, T., and Avouris, P. (1998) Single- and multiwall carbon nanotube field-effect transistors. *Appl. Phys. Lett.* **73**, 2447–2449.
4. Collins, P. G., Bradley, K., Ishigami, M., and Zetti, A. (2000) Extreme oxygen sensitivity of electronic properties of carbon nanotubes. *Science* **287**, 1801–1804.
5. Bradley, K., Jhi, S. H., Collins, P. G., Hone, J., Cohen, M. L., Louie, S. G., and Zettl, A. (2002) Is the intrinsic thermoelectric power of carbon nanotubes positive? *Phys. Rev. Lett.* **85**, 4361–4364.
6. Kong, J., Franklin, N. R., Zhou, C., Chapline, M. G., Peng, S., Cho, K., and Dai, H. (2000) Nanotube molecular wires as chemical sensors. *Science* **287**, 622–625.
7. Star, A., Gabriel, J. C. P., Bradley, K., and Gruner, G. (2003) Electronic detection of specific protein binding using nanotube FET devices. *Nano Lett.* **3**, 459–463.
8. Artyukhin, B. A., Stadermann, M., Friddle, R. W., Stroeve, P., Bakajin, O., and Noy, A. (2006) Controlled electrostatic gating of carbon nanotube FET devices. *Nano Lett.* **6**, 2080–2085.
9. Burghard, M., and Balasubramanian, K. (2005) Chemically functionalized carbon nanotubes. *Small* **1**, 180–192.
10. Chen, R. J., Zhang, Y., Wang, D., and Dai, H. (2001) Noncovalent sidewall functionalization of single-walled carbon nanotubes for protein immobilization. *J. Am. Chem. Soc.* **123**, 3838–3839.
11. Shim, M., Kam, N. W. S., Chen, R. J., Li, Y., and Dai, H. (2002) Functionalization of carbon nanotubes for biocompatibility and biomolecular recognition. *Nano Lett.* **2**, 285–288.
12. Besteman, K., Lee, J. O., Wiertz, F. G. M., Heering, H. A., and Dekker, C. (2003) Enzyme-coated carbon nanotubes as single-molecule biosensors. *Nano Lett.* **3**, 727–730.
13. Zhao, H., Asai, S., Kohno, T., and Ishikawa, K. (1998) Effects of brain temperature on CBF thresholds for extracellular glutamate release and reuptake in the striatum in a rat model of graded global ischemia. *Neuroreport* **9**, 3183–3188.
14. Hossmann, K. A. (1994) Viability thresholds and the penumbra of focal ischemia. *Ann. Neurol.* **36**, 557–565.
15. Busto, R., Globus, M. Y., Dietrich, W. D., Martinez, E., Valdes, I., and Ginsberg, M. D. (1989) Effect of mild hypothermia on ischemia-induced release of neurotransmitters and free fatty acids in rat brain. *Stroke* **20**, 904–910.
16. Benveniste, H., Drejer, J., Schousboe, A., and Diemer, N. H. (1984) Elevation of the extracellular concentrations of glutamate and aspartate in rat hippocampus during transient cerebral ischemia monitored by intracerebral microdialysis. *J. Neurochem.* **43**, 1369–1374.
17. Wahl, F., Obrenovitch, T. P., Hardy, A. M., Plotkine, M., Boulu, R., and Symon, L. (1994) Extracellular glutamate during focal cerebral ischaemia in rats: Time course and calcium dependency. *J. Neurochem.* **63**, 1003–1011.
18. Obrenovitch, T. P., Urenjak, J., Richards, D. A., Ueda, Y., Curzon, G., and Symon, L. (1993) Extracellular neuroactive amino

acids in the rat striatum during ischaemia: Comparison between penumbral conditions and ischaemia with sustained anoxic depolarization. *J. Neurochem.* **61**, 178–186.

19. Takagi, K., Ginsberg, M. D., Globus, M. Y., Dietrich, W. D., Martinez, E., and Kraydieh, S. (1993) Changes in amino acid neurotransmitters and cerebral blood flow in the ischemic penumbral region following middle cerebral artery occlusion in the rat: Correlation with histopathology. *J. Cerebr. Blood Flow Metab.* **13**, 575–585.
20. Hirota, H., Katayama, Y., Kawamata, T., Kano, T., and Tsubokawa, T. (1995) Inhibition of the high-affinity glutamate uptake system facilitates the massive potassium flux during cerebral ischaemia in vivo. *Neurol. Res.* **17**, 94–96.
21. Katayama, Y., Kawamata, T., Tamura, T., Hovda, D. A., Becker, D. P., and Tsubokawa, T. (1991) Calcium-dependent glutamate release concomitant with massive potassium flux during cerebral ischemia in vivo. *Brain Res.* **558**, 136–140.
22. Albery, W. J., Boutelle, M. G., and Galley, P. T. (1992) The dialysis electrode – A new method for in vivo monitoring. *J. Chem. Soc. Chem. Commun.* **12**, 900–901.
23. Lee, M., Im, J., Lee, B. Y., Myung, S., Kang, J., Huang, L., Kwon, Y. K., and Hong, S. (2006) Linker-free directed assembly of high-performance integrated devices based on nanotubes and nanowires. *Nat. Nanotechnol.* **1**, 66–71.
24. Lee, G. J., Lim, J. E., Park, J. H., Choi, S. K., Hong, S., and Park, H. K. (2009) Neurotransmitter detection by enzyme-immobilized CNT-FET. *Curr. Appl. Phys.* **9**, S25–S28.
25. Choi, S. K., Lee, G. J., Lim, J. E., Eo, Y. H., Kang, S. W., Han, J. H., Kim, B. S., Oh, B. S., and Park, H. K. (2008) Development of animal model for cerebral ischemia. *Korean J. Integr. Med.* **3**, 11–16.
26. Caragine, L. P., Park, H. K., Diaz, F. G., and And Phillis, J. W. (1998) Real-time measurement of ischemia-evoked glutamate release in the cerebral cortex of four and eleven vessel rat occlusion models. *Brain Res.* **793**, 255–264.
27. Lee, G. J., Choi, S. K., Eo, Y. H., Kang, S. W., Choi, S., Park, J. H., Lim, J. E., Hong, K. W., Jin, H. S., Oh, B. S., and Park, H. K. (2009) The effect of extracellular glutamate release on repetitive transient ischemic injury in global ischemia model. *Korean J. Physiol. Pharmacol.* **13**, 23–26.

Chapter 7

Kinesin I ATPase Manipulates Biohybrids Formed from Tubulin and Carbon Nanotubes

Cerasela Zoica Dinu, Shyam Sundhar Bale, and Jonathan S. Dordick

Abstract

This chapter describes a method for the formation of novel protein–nanotube hybrid conjugates. Specifically, we took advantage of the self-assembly and self-recognition properties of tubulin cytoskeletal protein immobilized onto carbon nanotubes to form nanotube-based biohybrids. Further biohybrid hierarchical integration in assemblies enabled molecular-level manipulation on engineered surfaces, as demonstrated with biocatalyst kinesin 1 ATPase molecular motor. The method presented herein can be extended for the preparation of biocatalyst-based or protein-based assemblies to be used as sensors or biological templates for nanofabrication.

Key words: Tubulin, carbon nanotube, self-assembly, self-recognition, biohybrid, kinesin-based manipulation.

1. Introduction

An exciting avenue in nanotechnology entails finding novel means to design and synthesize organic–inorganic hybrid materials with novel functions and properties that are not achievable in the individual components taken alone. For instance, by integrating biomolecules (e.g., proteins/enzymes, antigens/antibodies, or DNA/RNA) (1–3) with carbon nanotubes, biohybrid conjugates that combine the physical properties of the nanotube with the recognition or catalytic properties of the biomolecule were obtained (4). Such hybrid structures depend on either the native biomolecule function or the physicochemical properties of the template to achieve specific applications ranging from electronics

P. Wang (ed.), *Nanoscale Biocatalysis*, Methods in Molecular Biology 743,
DOI 10.1007/978-1-61779-132-1_7,

(5), to sensing elements (6), or antifouling materials (7). However, grand challenges reside in improved methods for hybrid separation and characterization and their hierarchical integration into functional assemblies to allow manipulation (8, 9).

In this chapter, we show that by harnessing the self-assembly and self-recognition properties of tubulin protein immobilized onto carbon nanotube, we form functional biohybrids that are manipulated in synthetic (non-physiological) environment by kinesin 1 molecular motors. Carbon nanotubes are allotropes of carbon with a cylindrical nanostructure (10) and novel properties that make them ideal candidates in applications including chemical nanowires (11), conductive films (12), bulb filaments (13), and optical ignition (14). Tubulin belongs to the small family of eukaryotic globular proteins; the most common members of the tubulin family are α- and β-tubulin. In the cell, α- and β-tubulin self-assemble in the presence of guanosine triphosphate (GTP) and in a head-to-tail fashion to form a protofilament (15). When 13 protofilaments associate laterally, they shape two-dimensional arrays that twist into a hollow cylindrical lattice called the microtubule (16). Microtubules are cytoskeletal filaments that serve as tracks for kinesin molecular motors loaded with different organelles (15, 17). The reason is that the eukaryotic cell cytosol is crowded and diffusion is too slow to efficiently assure transport of material from one part of a cell to another (15). Motors circumvent this problem by using adenosine triphosphate (ATP) as a fuel to produce directed movement along the cytoskeletal filaments. Kinesin 1 is a well-studied motor protein (18) with specific roles in the cell, including centrifugal movement and placement of lysosomes (19), mitochondrial motility (20), or axonal transport (21).

In order to form functional biohybrids, we used physisorption of tubulin onto acid-oxidized, multi-walled carbon nanotubes (MWNTs). The morphology of the resulting tubulin–MWNT conjugates was elucidated using atomic force microscopy (AFM) (**Fig. 7.1**). Specifically, we showed that tubulin attached onto the nanotube template; upon the tubulin concentration being used in the immobilization process and at a constant concentration of MWNTs, we observed a wide range of conjugate geometries from petal- to flower-like conformation. The interplay between the hydrophobic nature of tubulin, nanotube hydrophobic walls, the hydrophilicity of the mica surface, and the surface drying effects is thought to be responsible for the observed geometries. Further, in the presence of free tubulin and GTP in solution, the tubulin–MWNT conjugates served as templates for further tubulin polymerization. The tubulin polymerized onto the nanotube template led to hybrid microtubular structures that encapsulated aligned carbon nanotubes as confirmed by electron (**Fig. 7.2**) and fluorescence microscopy (**Fig. 7.3**).

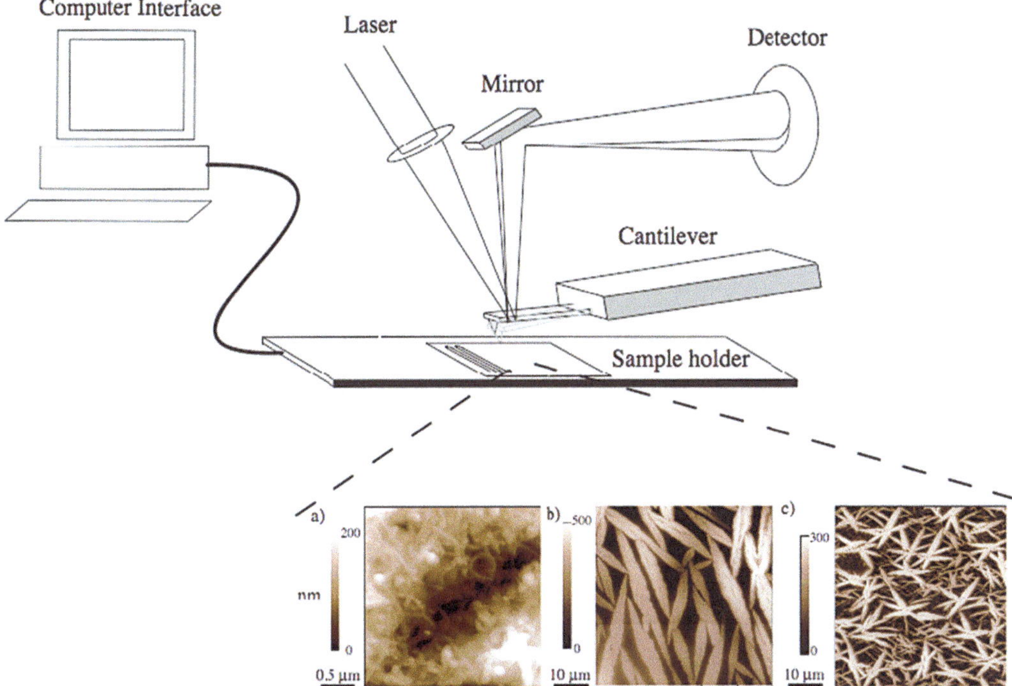

Fig. 7.1. Contact mode AFM of (**a**) MWNTs; (**b**) tubulin–MWNT conjugates formed by incubating 10 μM tubulin with 1 mg MWNTs; and (**c**) tubulin–MWNT conjugates formed by incubating 100 μM tubulin with 1 mg MWNTs. The hydrophobic/hydrophilic tubulin adsorption onto nanotubes, the adsorption of the conjugates onto the mica surface and the surface drying effects led to (**b**) petal-like or (**c**) flower-like conformation. The height was *color coded*, as indicated by the *vertical bars*. Reproduced with permission from Dinu et al. (22). Copyright Wiley-VCH Verlag GmbH & Co. KGaA.

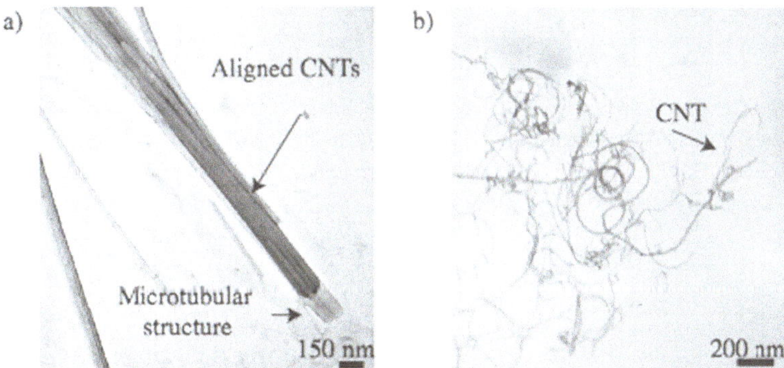

Fig. 7.2. (**a**) TEM images of linear biohybrids formed by free tubulin (5 μM) polymerization onto tubulin–MWNT conjugates (100 nM) in the presence of GTP, deposited onto carbon-coated formvar grids and stained with uranyl acetate. The extended microtubular structure encapsulates bundles of nanotubes as indicated by the *arrow*. (**b**) MWNTs deposited on a holey-carbon coated copper grid. Reproduced with permission from Dinu et al. (2009, *Small*). Copyright Wiley-VCH Verlag GmbH & Co. KGaA.

The functional properties of the biohybrids were subsequently exploited in a biological context by demonstrating manipulation in synthetic environment when glass-adsorbed kinesin 1 molecular motors and ATP was used (**Fig. 7.3**). Thus, the observed

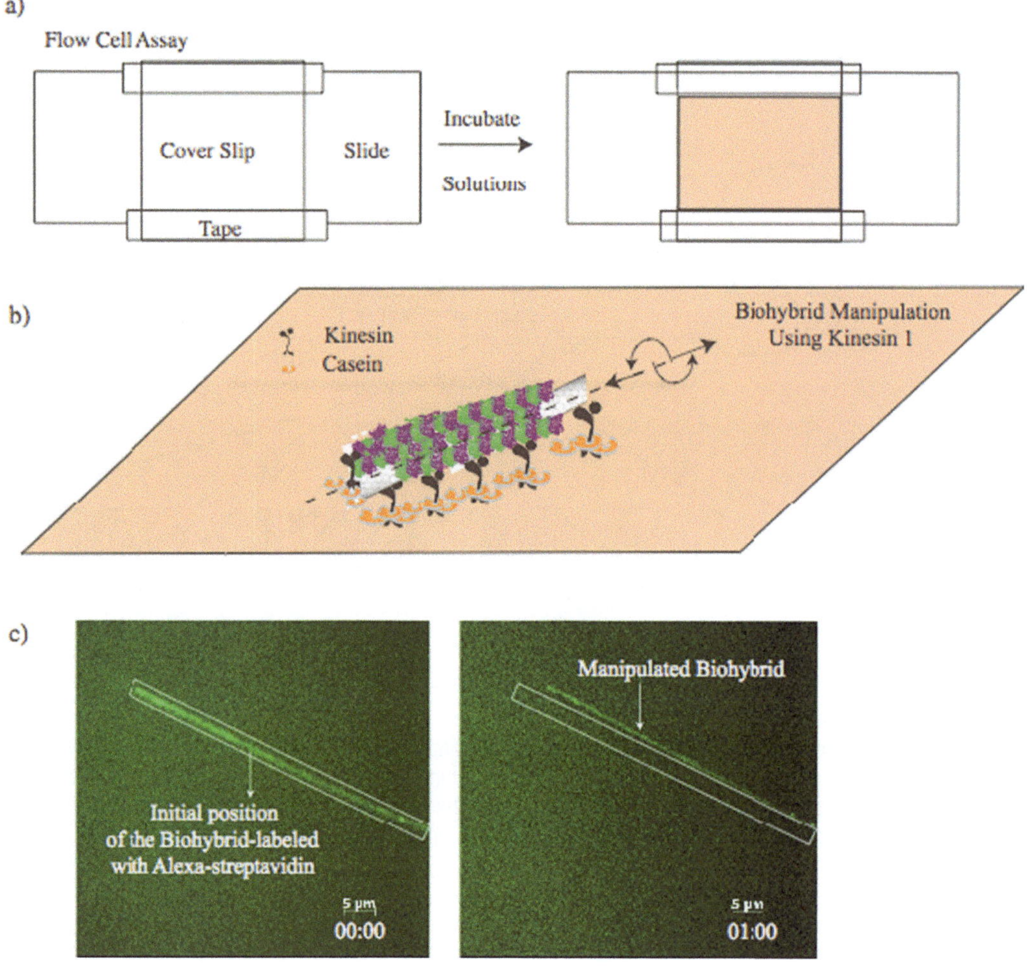

Fig. 7.3. Functional properties of biohybrids. (**a**) Schematic representation of a perfusion chamber. (**b**) Schematic representation of biohybrid manipulation on kinesin-coated surfaces. (**c**) Time-dependent analysis of biohybrid in fluorescence microscopy. The biohybrid translocates on kinesin-coated surface. Reproduced with permission from Dinu et al. (22). Copyright Wiley-VCH Verlag GmbH & Co. KGaA.

polymerization based on biomolecular recognition and intrinsic organizational self-assembly of tubulin yielded highly functional biohybrids with potential applications as sensors, biological templates for nanofabrication, and building blocks for hierarchical supramolecular assembly.

2. Materials

2.1. Tubulin

1. Tubulin (>99% pure) purified from bovine brain using a method by Shelanski et al. (23) was purchased from Cytoskeleton, Inc., USA, in lyophilized form.

2. Biotin-labeled tubulin (>99% pure) obtained by modifying the lysines on the tubulin surface to randomly contain a covalently linked, long chain of biotin derivative was supplied as a lyophilized powder (Cytoskeleton, Inc., USA).
3. Rhodamine-labeled tubulin (>99% pure) modified to contain covalently linked rhodamine at random surface lysines was purchased from Cytoskeleton, Inc., USA, and supplied as lyophilized powder. Rhodamine-labeled tubulin is detected using a filter set of 535 nm excitation and 585 nm emission.
4. Upon arrival, the samples should be stored at −80°C to avoid protein degradation.

2.2. Multi-walled Carbon Nanotubes (MWNTs)

MWNTs produced by chemical vapor deposition (CVD) to strict lengths (1–5 μm) and diameter specifications (15±5 nm) and purified to >95% as measured by thermogravimetric analysis (TGA) were purchased from NanoLab, Newton, MA, USA. The residuals or catalysts in MWNTs are identified by EDAX and may include iron and sulfur. The specific surface area (SSA) of the MWNTs was estimated by the manufacturer at 200–400 m^2/g. The nanotubes are stored at room temperature.

2.3. BRB 80 Working Buffer Preparation

Weigh PIPES (piperazine-1,4-bis(2-ethanesulfonic acid); Sigma, USA) to lead to a final concentration of 80 mM in 1 L final volume. Weigh corresponding amounts to give 1 mM $MgCl_2$ (magnesium chloride; Sigma, USA), 1 mM EGTA (ethylene glycol-bis (2-aminoethylether)-*N, N, N′,N′*-tetra acetic acid; Sigma, USA), and 100 mM KOH (potassium hydroxide pellets; Sigma, USA) in 1 L final volume. Place all powders in a 1-L glassware bottle (Fisher, USA), add 800 mL Milli-Q water, and stir overnight. Carefully add KOH pellets until the buffer solution is totally clear; adjust the pH to 6.8 and fill to 1 L. Filter through Millipore Stericups filter units, 0.22 μm (Millipore, USA), aliquot in Falcon tubes (VWR, USA), and store at −20°C.

2.4. Uranyl Acetate Preparation

Prepare 0.5% uranyl acetate solution by dissolving 50 mg uranyl acetate (Electron Microscopy Sciences, USA) in 10 mL Milli-Q water. Uranyl acetate is radioactive and has to be handled carefully under proper supervision and lab settings. Shake the solution at 200 rpm at 4°C until it is fully dissolved. Filter the solution through a 0.22-μm filter and store in an appropriate dark glass container in a designated radioactive area. Holey carbon-coated copper grids, 200 mesh, and formvar-coated copper grids, 400 mesh (Electron Microscopy Sciences, USA), are being used for imaging MWNTs and tubulin–MWNT conjugates, respectively.

2.5. Prepare the Microscopy Solutions

1. Weigh dry powder of GTP (Sigma, USA) and dissolve it into the appropriate volume of Milli-Q water to lead to a 25 mM GTP concentration. Prepare aliquots, quick-freeze in liquid nitrogen, and store at −20°C.
2. Weigh dry powder of $MgCl_2$ (Sigma, USA) and dissolve it into the appropriate volume of Milli-Q water to lead to a 100 mM concentration. Prepare aliquots, quick-freeze in liquid nitrogen, and store at −20°C.
3. Weigh dry powder of paclitaxel (Sigma, USA) and dissolve it into the appropriate volume of DMSO (dimethyl sulfoxide; Sigma, USA) to lead to 1 mM concentration paclitaxel. Make aliquots, quick-freeze in liquid nitrogen, and store at −20°C.
4. Weigh dry powder of ATP (Sigma, USA) and dissolve into appropriate volume of 100 mM $MgCl_2$ to give a theoretical concentration of 100 mM. Dilute 1:1000 in Milli-Q water and check the pH. Make aliquots, quick-freeze in liquid nitrogen, and store at −20°C.
5. Prepare stock solution of 10 mg/mL casein (Sigma, USA) in BRB 80. Using a 50-mL Falcon tube, dissolve 1 g dry casein in ~35 mL BRB 80. Place in a tumbler in the cold room and allow large precipitates to settle overnight. Take supernatant and spin it at 5,000 rpm to pellet out any remaining precipitates. Carefully remove supernatant and filter through 0.2-μm Millipore filter membrane. Measure absorbance at 280 nm with blank against BRB 80. Correct dilution for desired concentration; aliquot and quick-freeze in liquid nitrogen.
6. Prepare the casein microscopy solution by diluting casein to a final concentration of 0.5 mg/mL in BRB 80.
7. Prepare the kinesin microscopy solution by diluting the kinesin to a final concentration of 10 μg/mL in BRB 80 containing 0.1 mg/mL casein and 1 mM ATP. Wild-type kinesin-1 (full-length drosophila expressed in bacteria and purified as described by Coy et al. (24)) was used.

2.6. Perfusion Chamber

1. Use a clean microscope slide (Fishersci, USA); stick two pieces of double-sided tape on each of its larger sides (Scotch 3 M, thickness 0.1 mm; Staples, USA). The tapes are separated by at least 5 mm (**Fig. 7.3**).
2. Press a cover slip (Corning cover slip, 22×22, 24×24, 24×50 mm^2; Corning, USA) on top of the double-sided tape.

3. Methods

3.1. Prepare Rhodamine-Labeled Biotinylated Tubulin

1. Suspend the lyophilized powders of rhodamine-labeled tubulin, unlabeled tubulin, and biotin-labeled tubulin in the appropriate volume of BRB 80 (see the manufacturer's notes) and store in an ice-containing box to lead to 4 mg/mL final concentration of tubulin.
2. Prepare rhodamine-labeled biotinylated tubulin (final concentration 4 mg/mL) by mixing rhodamine-labeled tubulin, biotinylated tubulin, and unlabeled tubulin in a ratio of 3:4:9.
3. Aliquot the rhodamine-labeled biotinylated tubulin (10 μL solution in 250 μL aliquot; VWR, USA). Freeze the aliquots in liquid nitrogen and store at −80°C until use.

3.2. MWNT Oxidation

1. Weigh 100 mg of MWNTs and place the nanotubes in a 1-L Erlenmeyer flask (WVR, USA). The nanotube powder should be handled with care; use a mask to avoid nanomaterial inhalation.
2. Suspend the MWNTs in acid mixture containing H_2SO_4 (sulfuric acid; Fisher, USA) and HNO_3 (nitric acid, 90%; Fisher, USA) in a ratio of 3:1 (v/v). Mix 60 mL acid mixture with 100 mg MWNTs under a fume hood. The acid mixture is highly corrosive, thus it should be handled with care. Wear lab glasses, lab coats, and avoid contact with skin.
3. Sonicate the nanotubes–acid mixture for 3 h in a water bath sonicator (Model 50T; VWR International, USA) at a rated power of 45 W. To prevent overheating the acid–nanotube mixture, periodically replace the water with ice. Oxidation leads to cut, acid-functionalized MWNTs that are highly dispersible in water. Longer cutting times lead to shorter nanotubes, while an increase in temperature accentuates the cutting efficiency.
4. After 3 h, remove the oxidation solution (600 mL) from the sonicator and dilute it gradually with constant swirling into an Erlenmeyer flask (VWR, USA) containing 900 mL of ice-cold Milli-Q water. Allow 10 min for heat dissipation. Prepare the filtering unit (Millipore, USA). All the experiments are to be performed in the hood and with proper lab protection equipment.
5. Filter the diluted nanotube solution through the Millipore 0.2-μm polycarbonate filter membrane GTTP (isopore membrane; Millipore, USA). Wash extensively with Milli-Q to remove any residues. Re-disperse the mass of the MWNTs

from the filter paper in fresh 1,000 mL Milli-Q water and repeat the filtration process at least four times (every time use a new filter membrane) until water-soluble MWNTs are obtained; the pH of the filtrate should be neutral. To remove the nanotubes from the filter paper, sonicate the filter in Milli-Q water for 5–10 min or until all the nanotubes are dispersed.

6. Dry and remove the oxidized and thus cut nanotubes from the filter paper and store them at room temperature in a fresh glass vial (20 mL; VWR, USA).

3.3. Prepare Tubulin–MWNT Conjugates

Tubulin–MWNT conjugates are prepared by physical adsorption of tubulin onto MWNT nanotubes:

1. Prepare 1 mL fresh solution of given concentration of tubulin in BRB 80. The tubulin concentration is chosen to assure either partial or full coverage of the nanotube based on the available SSA as listed by the manufacturer.
2. Resuspend 1 mg of oxidized MWNTs in the 1 mL tubulin solution.
3. Incubate the nanotube–tubulin mixture at 4°C for 2 h with shaking at 200 rpm on a platform shaker (IKA HS 260, USA). Conjugates were formed from 10 μM tubulin immobilized on 1 mg MWNTs (for petal-like conformation) and 100 μM (for flower-like conformation).
4. After incubation, disperse the resulting tubulin–MWNT conjugates into an Eppendorf microcentrifuge tube (1.5 mL; VWR, USA) and centrifuge at 10,000 rpm for 3 min in a centrifuge with swinging bucket rotor (LabNet, Hermle Z 233 M-2, USA).
5. Carefully decant and collect the supernatant using a pipettor (avoid loss of the conjugates). Resuspend the settled conjugates in BRB 80 by tipping gently and inverting the tube. Repeat the centrifuging and washing steps at least five times to remove any loosely bound tubulin. Collect all supernatant fractions. Alternatively, the conjugates can be washed and supernatant collected using the filtering unit as previously described for MWNT oxidation.
6. Disperse the 1 mg washed conjugates in 1 mL fresh BRB 80 buffer to reach 1 mg/mL conjugate concentration.

3.4. BCA Protein Assay

Protein content (defined as protein loading on the MWNT) can be measured in the range of 5–2,000 μg/mL using the commercial BCA protein assay kit (Pierce, USA). This assay kit contains two reagents: reagent A, containing sodium carbonate, sodium bicarbonate, bicinchoninic acid (BCA), and sodium tartrate in 0.1 M sodium hydroxide, and reagent B, containing 4% cupric sulfate. The assay combines the reduction of Cu^{2+} to Cu^{1+} by protein in an alkaline medium with the highly sensitive and selective colorimetric detection of the cuprous cation (Cu^{1+}) by BCA. The BCA/copper complex is water soluble and exhibits a strong linear absorbance at 562 nm as a function of protein concentrations. The assay is described below

1. Mix 50 parts of reagent A (1,000 μL) with 1 part of reagent B (20 μL) (50:1 reagent A:reagent B) to form the working reagent (WR). Prepare sufficient volume of WR based on the number of samples to be assayed (number of the supernatants collected as previously described). Vortex. When reagent B is first added to reagent A, a turbid green color is observed; the sample becomes clear once the solution is vortexed.
2. In an appropriately labeled aliquot of 1.5 mL, add 50 μL of sample and 1,000 μL of the WR. Mix well. Prepare a control with 50 μL of BRB 80 in 1,000 μL of WR.
3. Incubate the aliquots at 60°C for 30 min. The incubation period was chosen to yield maximal reaction in a reasonable time frame. After incubation, cool down the samples.
4. Use fresh 1.5-mL semi-microdisposable cuvette (VWR, USA) and pipette 1,000 μL of the BCA assay mixture.
5. Measure the absorbance at 562 nm on a UV–Vis spectrophotometer (Shimadzu UV-2401, Columbia, MD, USA). Make the baseline using the control, i.e., buffer–BCA reagent mixture. Continue measuring the samples and record the absorbance values.
6. Determine the amount of protein attached onto the MWNTs by the difference of the initial amount of tubulin added and the amount of tubulin washed out in the supernatant. Standard calibration curves should be prepared using the corresponding purified protein in the BRB 80.

3.5. Sample Preparation for AFM

1. Prepare an epoxy mixture by vigorously mixing 1 mL of resin with 1 mL hardener (Devcon 5-min epoxy; Danvers, MA, USA).
2. Attach the mica sheet (AFM-grade muscovite mica sheets; Electron Microscopy Sciences, USA) onto a clean microscope slide (corning, USA) using epoxy. Specifically, spot a drop of freshly mixed epoxy on a clean glass slide and press the mica sheet firmly using cotton swabs (VWR, USA). Ensure that the mica sheet is attached properly on all the sides and is relatively flat. Avoid excess epoxy protruding from the edges as it might interfere with the AFM scan head.
3. Before spotting, freshly remove the top layers of the mica sheets using scotch tape to assure exposure of a clean,

unused surface. In short, press a fresh strip of scotch tape onto the mica and gently pull off the top layer. Repeat this procedure till the topmost exposed layer is flat, without residue or cracks. Verify by looking at the mica sheet at an angle. The mica sheets can be reused for new samples by cleaving any top layers that have been previously used or exposed.

4. Dilute and disperse the conjugates to a final concentration of 0.1 mg/mL of carbon nanotube in BRB 80. Spot ~10 μL sample onto the freshly cleaved mica using a pipettor.
5. Dry the samples overnight in a dessicator (Belart, USA) under vacuum. For better imaging, make sure that the samples are stored prior to imaging in a moisture-free environment to avoid image artifacts.

3.6. AFM Imaging

This is a brief protocol for imaging on Asylum Research Molecular Force Probe 3D (MFP-3D). The samples are imaged in the tapping mode (AC mode) (**Fig. 7.1**):

1. Before starting the equipment, ensure that the AFM scan head is properly connected to the controller.
2. Detach the cantilever holder from the scan head and carefully insert a fresh cantilever (AC-160 TS; Asylum Research, USA).
3. Adjust the cantilever with the polished surface on the back to accommodate the laser. For this, move the cantilever 3–4 mm below the edge of the polished area. Tighten the screw to hold the cantilever firmly in place.
4. Insert the cantilever holder into the scan head of the AFM machine and start the controller and the software (MFP-3D). The software is developed on IgorPro, an open source program, and is easily customizable to work with the AFM.
5. Align the laser with the tip. For this, visualize the laser using an infrared sensor card (Newport, Irvine, CA). Adjust the laser to graze the bottom of the tip in the cantilever and move the laser along the edge (X-direction) until the cantilever obstructs the laser. Once the base of the cantilever is identified, move the laser in the Y-direction (outward to the end of the tip).
6. Adjust the laser to obtain the maximum signal by moving it to the very end of the tip; follow the signal on the computer software until a SUM (where the SUM is the total laser signal recorded at the detector) of 6–7 is reached.
7. Adjust the deflection to read zero. This ensures capture of any deviations in the laser reflection in both positive and negative directions.

8. Once the laser is aligned onto the tip, mount the sample on the AFM scanner and hold in place using the provided magnets. Ensure that the AFM scan head is raised to its maximum height to avoid the cantilever crashing into the sample.
9. Using the MFP-3D software, tune the cantilever to find the first harmonic frequency. This varies with the type of cantilever used; an estimate should be provided by the supplier, e.g., for AC-160 TS, the first harmonic frequency is around 300 kHz. For auto-tune, set the scan to include the recommended frequency (usual range is 40–350 kHz).
10. Adjust the drive frequency to –5% of the harmonic frequency obtained. The amplitude should read 1.0 V.
11. Once the cantilever is tuned, set an image scan rate (usually at 1 Hz), set point (800 mV), scan lines and points (256 or 512) and save location. The cantilever is not yet in contact with the sample surface.
12. Engage the cantilever. This tilts the tip toward the sample and changes the Z-piezo voltage to 150.0.
13. Slowly lower the scan head manually by simultaneously manipulating its three legs; notice the changes in the amplitude value. *Do not* crash the tip onto the sample.
14. As the tip approaches the surface, the Z-piezo voltage decreases from 150.0 to 70.0. Stop lowering the head once it reaches 70.0. At this point, the amplitude reads 800 mV, which equalizes the chosen set point.
15. Start scanning the surface by choosing "Do Scan," "Frame Down," or "Frame Up" (this defines the scanning direction). The resulting image will have three profiles: the height trace, the phase trace, and the amplitude.
16. Select the area of interest in the scanned image and "fix scale." This adjusts the data (the zero for an AFM is not defined).
17. Adjust the set point and the integral gain. In particular, adjust the set point (usually decrease it) and increase integral gain to match both the trace and the re-trace for the height scan. After adjusting the set point and integral gain, re-start scanning the image.
18. Process the image by flattening and sharpening its features. Specifically, flatten the image by zero order using the MODIFY panel of the software. Select the features of interest using an iterative selection process in the mask tab. This usually selects the points in the image that are "lower" than the "features." It is important to "inverse" the selection (which as a result will include the higher areas of interest

in the image) and check "fill" the selection to protect its features. Since the edges of the features are very important for estimating its physical dimensions, it is advisable to "dilate" the mask two times and "erode" it once. Once a suitable mask is found, flatten the rest of the image (which is predominantly mica) using first-order flattening function in the MODIFY tab. The processed image is now ready for analysis which can include, e.g., the height of features and scan-line data profile.

19. The flattened images are rendered using a "desert solitude" palette. The color bar representing the *Z*-axis (height) is adjusted using height offset and scale to reflect all the captured features and exported as TIFF image files with 300 dpi.

3.7. TEM Sample Preparation

1. In a fresh aliquot, dilute the sample (tubulin–MWNT conjugates or oxidized carbon nanotubes when controls are performed) in BRB 80 to a final concentration of 0.1 mg/mL. Disperse the sample by pipetting; keep the aliquot handy.
2. Place the TEM grid on a filter paper (to remove any additional fluid; use Whatman filter paper, e.g., with diameter of 5.5 cm; VWR, USA) in a clean Petri dish (VWR, USA) with the help of a pair of tweezers. Maintain the coated side (holey carbon or formvar) on the top. The coated side of the grid can be visualized by contrast difference. The grid is very sensitive to any static charge, thus it may need to be rearranged to accommodate the solution.
3. Draw sample solution and gently spot it onto the grid. This technique is generally referred to as "drop casting." (a) For imaging the nanotubes, spot 10 μL of the sample on the carbon-coated grid; (b) for imaging of the tubulin–MWNT conjugates, place 3–5 μL of the conjugates onto the formvar side of the grid and allow to settle for about 3 s. Subsequently add 10–15 μL of 0.5% uranyl acetate solution and wait for about 3 s. Drain the excess liquid by gently touching it with the filter paper. For better results, move the grid to a fresh filter paper.
4. Dry the samples overnight in a dessicator under vacuum. Store the samples in a vacuum-tight container and in a dry place prior to imaging.

3.8. TEM Imaging

This brief protocol is for TEM imaging on Philips CM-12 transmission electron microscope (**Fig. 7.2**):

1. Before starting the machine, check the vacuum and make sure it reads READY; fill the anti-contamination device (ACD) with liquid nitrogen.

2. Apply high tension in steps of 20 kV to a maximum tension of 120 kV.
3. Check vacuum status at 80, 100, and 120 kV and ensure it is in READY mode. If otherwise, step back on the voltage and wait for vacuum stabilization before further increasing the voltage.
4. Once the machine is stable at 120 kV, increase the filament current to 16–18 μA in steps of 1 μA and at 30 s intervals until an image of the filament appears on the machine screen.
5. At 16–18 μA, adjust the condenser stigmators by focusing on the filament image. Place the condenser stigmators in STIG page; focus to obtain a sharp filament image. Bring the filament to full current (usually notified by the manager/operator function) and center the condenser aperture manually by expanding the spot (using Intensity button). Align the spot with the screen.
6. Adjust gun TILT in the ALIGN page to obtain the maximum brightness.
7. Align the column by following the instructions in the ALIGN-GUN page. This is a six-step procedure for beam alignment and for the various spot sizes of the column. Once the beam is aligned at all spot sizes, adjust the stigmators to obtain a circular spot with minimum deviation of the spot with varying intensity, as described in Step 3.
8. Bring the spot size to 2 (the spot size where imaging is done) and expand to cover the viewing screen.
9. Remove the sample holder and mount the grid with the sample side facing up. Handle the grid very carefully using a pair of sharp, clean tweezers.
10. Obtain the image of the sample by adjusting the magnification and the intensity of the beam. View the samples in the "bright-field mode" of the TEM. For typical imaging of carbon nanotubes, use a magnification of 100,000 or above. For sharp image with good contrast, focus on one of the edges of the sample to remove any diffraction fringes. To enhance the contrast, insert the objective aperture prior to collecting the image. Capture the images and record on film. For good quality images, adjust the intensity of the beam to a 4 s exposure time.
11. After capturing the images, remove the negatives from the TEM by purging the camera with nitrogen followed by releasing the vacuum. Ensure that the case holding the negatives is properly sealed.

12. The negatives are handled in a dark room. Mount negatives on a rack and immerse in a developer solution (Kodak, USA) for 3–4 min. Wash the developed negatives in running water for 2 min and transfer to a fixer solution (Kodak, USA) for 2–3 min. Then wash the fixed negatives in running water for 15 min, remove, and dry overnight. Wear gloves to avoid prints on the images. To ensure that negatives are fully immersed in fixer and developer, remove and re-immerse the negative rack once every 30 s.
13. Scan TEM negatives at 600 dpi and process.

3.9. Polymerize Tubulin–MWNTs with Free Tubulin Leading to Biohybrids

1. Mix conjugates of tubulin–MWNTs (obtained as previously described) in BRB 80 with free tubulin (5 μM).
2. Add the polymerization cocktail mixture consisting of 4 mM $MgCl_2$, 1 mM Mg-GTP, and 5% DMSO final concentration. Incubate solution at 37°C for 30 min.
3. The tubulin polymerized onto the nanotube template forms a biohybrid. Stabilize and dilute 100-fold at room temperature in BRB80 containing 10 μM paclitaxel.
4. Use 50 μL biohybrid solution and stain with Alexa-Fluor 488-labeled streptavidin (Invitrogen, USA) of 100 nM final concentration in BRB 80 and containing 10 μM taxol. The streptavidin binds specifically to the biotinylated tubulin that polymerized.

3.10. Microscopy Assay

1. In the perfusion chamber, perfuse 30 μL of the casein microscopy solution. Incubate at room temperature for 5 min to allow protein binding. Casein is used to prevent non-specific binding of the kinesin or biohybrids to the surface.
2. Perfuse 30 μL of the kinesin microscopy solution and allow protein binding for 5 min.
3. Perfuse 30 μL of the motility solution containing the stained biohybrids (in this case, the biohybrids were formed from 10 nM tubulin immobilized on MWNTs and 5 μM free tubulin), 1 mM ATP, and 1 μM excess biotin (Sigma, USA) (**Fig. 7.3**).

3.11. Fluorescence Microscopy

This is a brief protocol for imaging on a Zeiss Axio Observer Z1 (fully motorized version, including motorized Z-drive). The modulus used for imaging, AxioVision, is connected to a high-definition AxioCam CCD camera, with variable exposure time ranging from 1 ms to 60 s, with up to 48 images/s and flexible resolutions up to 1,388 × 1,040 pixels (**Fig. 7.3**):

1. Switch on the mercury lamp (HBO 100). The lamp needs to be switched on about 5–10 min before use.

2. Switch on the power supply.
3. Switch on the Hamamatsu camera controller.
4. Switch on the microscope. As soon as the microscope is switched on, the microscope controller starts glowing. After about 15 s, the control becomes active.
5. Adjust the focus. Use the controller, hit 10× and illumination "on." Choose bright field.
6. Adjust the eyepiece – normal viewing condition would be aligning the "0" mark to the *yellow* dot on the base of the eyepiece (manufacturer's instructions). Adjust the viewing angle between the eyepieces according to what suits user eyes. After adjustment, a single perfect circle defined as your field of view should become visible.
7. Focus on the object using the coarse and fine focus knobs. Once the object seems to be in focus (sharper edges, etc.), adjust the field diaphragm to be completely closed (go to the farthest right).
8. Remove one of the eyepieces from its slot and look into the cavity. A circle or a hexagon should be visible. Increase or decrease the aperture so that the circle becomes the hexagon or if there is already a hexagon, change the aperture until all the sides of the hexagon are visible and located within the cavity. Place the eyepiece back.
9. Place the sample onto the sample holder so that the desired viewing area is directly above the objective lens. Use the stage control to locate this area.
10. Open AxioVision software. Set the imaging parameter (e.g., objectives, lighting path and variables, exposure parameters for the camera in use, the filters, and exposure time) using Axio Observer Z1. For picture acquisition, it is recommended to direct all the light to the camera. The exposure time should be kept low (<100 ms) for faster and smoother system operation and to avoid sample photobleaching during fluorescence imaging. Keep the hue% around 85 to capture the low brightness regions.
11. Control the Advanced Camera Tab. Adjust the camera settings (e.g., white balance and shading and gain).
12. Focus on the sample. You can choose either autofocus or manual focus mode. The system can automatically measure the working distance and optimal focus position.
13. Adjust the time lapse imaging parameters. Set the time step and specify the time interval between two consecutive images as well as the duration of the image acquisition.

14. Hit start.
15. Once the images are acquired, adjust the image properties to optimize its quality (e.g., brightness, contrast, and gamma value) using the characteristic curve of the software (i.e., recommended setting for gamma curve is 0.45).
16. Perform image analysis to obtain quantitative information (i.e., numerical data).

4. Notes

1. MWNTs can also be stored as aqueous solution in a recommended final concentration of 1 mg/mL.
2. Control experiments to prove the functionality of the biohybrids in a ligand recognition reaction can be performed in a microscopy perfusion chamber coated with antitubulin antibodies.
3. Control experiments performed in ethanol and ice (conditions that can disturb microtubule polymerization) did not affect the biohybrid stability and functionality.
4. The moving path of the biohybrid can be identified and compared with a moving path of a microtubule.
5. Kinesin 1 mechanical cycle for biohybrids can be identified and compared with the kinesin mechanical cycle for the microtubules.
6. Other types of carbon nanotubes (pristine and oxidized) can be used to study structural or functional changes and biohybrid behavior and functionality.

Acknowledgments

This work was supported by an NSF Nanoscale Science and Engineering Center (DMR 0642573). We thank Stefan Diez (Max-Planck Institute for Molecular Cell Biology and Genetics, Dresden, Germany) for providing kinesin and Douglas B. Chrisey (Department of Materials Science and Engineering, Rensselaer Polytechnic Institute) for the fluorescence microscopy facility.

References

1. Asuri, P., et al. (2006) Water-soluble carbon nanotube-enzyme conjugates as functional biocatalytic formulations. *Biotechnol. Bioeng.* **95**(5), 804–811.
2. Gazit, E. (2007) Use of biomolecular templates for the fabrication of metal nanowires. *FEBS J.* **274**(2), 317–322.
3. Lenihan, J. S., et al. (2004) Protein immobilization on carbon nanotubes through a molecular adapter. *J. Nanosci. Nanotechnol.* **4**(6), 600–604.
4. Cui, D. (2007) Advances and prospects on biomolecules functionalized carbon nanotubes. *J. Nanosci. Nanotechnol.* 7(4–5), 1298–1314.
5. Keren, K., et al. (2003) DNA-templated carbon nanotube field-effect transistor. *Science* **302**(5649), 1380–1382.
6. Besteman, K., et al. (2003) Enzyme-coated carbon nanotubes as single-molecule biosensors. *Nano Lett.* **3**(6), 727–730.
7. Stafslien, S. J., et al. (2007) Combinatorial materials research applied to the development of new surface coatings VI: An automated spinning water jet apparatus for the high-throughput characterization of fouling-release marine coatings. *Rev. Sci. Instrum.* **78**(7), 072204.
8. Bianco, A., and Prato, M. (2003) Can carbon nanotubes be considered useful tools for biological applications? *Adv. Mater.* **15**(20), 1765–1768.
9. Zhang, S. (2003) Fabrication of novel biomaterials through molecular self-assembly. *Nat. Biotechnol.* **21**(10), 1171–1178.
10. IIjima, S. (1991) Helical microtubules of graphitic carbon. *Nature* **354**, 56–58.
11. Hu, J., Odom, T. W., and Lieber, C. M. (1999) Chemistry and physics in one dimension: Synthesis and properties of nanowires and nanotubes. *Acc. Chem. Res.* **32**(5), 435–445.
12. Pei, S., et al. (2009) The fabrication of a carbon nanotube transparent conductive film by electrophoretic deposition and hot-pressing transfer. *Nanotechnology* **20**, 235707–235714.
13. Wei, J., Zhu, H., and Wu, D. (2004) Carbon nanotube filaments in household light bulbs. *Appl. Phys. Lett.* **84**, 4869–4872.
14. Hart, A. J., van Laake, L., and Slocum, A. H. (2007) Desktop growth of carbon-nanotube monoliths with in situ optical imaging. *Small* **3**(5), 772–777.
15. Dinu, C. Z., et al. (2007) Cellular motors for molecular manufacturing. *Anat. Rec. (Hoboken)* **290**(10), 1203–1212.
16. Nogales, E., and Wang, H. W. (2006) Structural intermediates in microtubule assembly and disassembly: How and why? *Curr. Opin. Cell Biol.* **18**(2), 179–184.
17. Zhang, Y., et al. (2007) Chemically encapsulated structural elements for probing the mechanical responses of biologically inspired systems. *Langmuir* **23**(15), 8129–8134.
18. Brady, S. T. (1985) A novel brain ATPase with properties expected for the fast axonal transport motor. *Nature* **317**(6032), 73–75.
19. Hollenbeck, P. J. (1990) Cell biology. Cytoskeleton on the move. *Nature* **343**(6257), 408–409.
20. Tanaka, Y., et al. (1998) Targeted disruption of mouse conventional kinesin heavy chain, kif5B, results in abnormal perinuclear clustering of mitochondria. *Cell* **93**(7), 1147–1158.
21. Muresan, V. (2000) One axon, many kinesins: What's the logic? *J. Neurocytol.* **29**(11–12), 799–818.
22. Dinu, C. Z., Bale, S. S., Zhu, G., and Dordick, J. S. (2009) Tubulin encapsulation of carbon nanotubes into functional hybrid assemblies. *Small*, **5**(3), 310–315.
23. Shelanski, M. L., Gaskin, F., and Cantor, C. R. (1973) Microtubule assembly in the absence of added nucleotides. *Proc. Natl. Acad. Sci. USA* **70**(3), 765–768.
24. Coy, D. L., Wagenbach, M., and Howard, J. (1999) Kinesin takes one 8-nm step for each ATP that it hydrolyzes. *J. Biol. Chem.* **274**(6), 3667–3671.

References

Chapter 8

Reversible His-Tagged Enzyme Immobilization on Functionalized Carbon Nanotubes as Nanoscale Biocatalyst

Liang Wang and Rongrong Jiang

Abstract

Common enzyme immobilization methods on nanomaterials (adsorption, covalent binding, crosslinking, encapsulation) often generate problems in enzyme leaching, 3D structure change and diffusion resistance. We show here a detailed site-specific enzyme immobilization method that overcomes the foresaid limitations. It is based on the specific interaction between His-tagged enzyme and single-walled carbon nanotubes modified with N_α,N_α-bis(carboxymethyl)-L-lysine hydrate. This method does not require enzyme purification and the resulting nanoscale biocatalyst can maintain high enzyme activity and stability. The enzyme-loading capacity is also comparable with the reported immobilization capacity on carbon nanotubes by either covalent binding or adsorption. Furthermore, the immobilization is reversible for several cycles while maintaining high enzyme activity and the nanoscale biocatalyst can be regenerated easily.

Key words: Single-walled carbon nanotubes, enzyme reloading, enzyme immobilization, nanomaterials, His-tagged enzyme.

1. Introduction

Enzyme immobilization on nanomaterials has attracted a lot of attention in research recently. Compared with bulk materials, nanomaterials can perform better during immobilization owing to their high surface area/volume ratio, good enzyme-loading capacity and low mass transfer resistance (1). However, common immobilization methods, including adsorption (2), covalent binding (3), entrapment (4), and encapsulation (5), often cause problems in enzyme leaching (6), enzyme 3D structure change (7), and diffusion resistance (8). The preservation of

P. Wang (ed.), *Nanoscale Biocatalysis*, Methods in Molecular Biology 743,
DOI 10.1007/978-1-61779-132-1_8,

enzyme activity is also low after immobilization, for example, carbon nanotube–enzyme immobilization by physical adsorption usually keeps 40–70% of the native enzyme activity (9), covalent attachment often maintains ~50% or less (10). Furthermore, all of these methods require pure enzyme for immobilization. Here in this work, we show a site-specific enzyme immobilization method that overcomes these limitations. We demonstrate an efficient immobilization method based on the specific interaction between His-tagged enzyme and single-walled carbon nanotubes (SWCNTs) modified with N_α,N_α-bis(carboxymethyl)-L-lysine hydrate (ANTA). This process does not require enzyme purification and the immobilization is reversible. The specific interaction between His-tagged enzyme and SWCNTs helps to keep enzyme 3D structure and orientation during catalysis, which results in high retention of enzyme activity. The His-tag also makes it possible to regenerate the SWCNT–enzyme conjugate.

2. Materials

2.1. Cell Culture and Lysis

1. Luria–Bertani (LB) broth for cell culture: 10 g tryptone, 10 g NaCl, and 5 g yeast extract are dissolved in 1 L distilled water, and autoclaved at 121°C for 25 min.
2. Isopropyl β-D-1-thiogalactopyranoside (IPTG) solution: 1 *M* of IPTG is dissolved in distilled water, filtrated by 0.2 μm filter, and stored at −20°C.
3. 4-(2-hydroxyethyl)-1-Piperazineethanesulfonic acid (HEPES) solution for cell lysis: 20 m*M* of HEPES is dissolved in distilled water at pH 7.5, filtrated by 0.2 μm filter, and stored at 4°C.

2.2. SDS-Polyacrylamide Gel Electrophoresis (SDS-PAGE)

1. Ammonium persulfate solution: 10% (w/v) of ammonium persulfate is dissolved in distilled water and stored at −20°C.
2. Sodium dodecyl sulfate solution (SDS): 10% (w/v) and 0.2% (w/v) of SDS are dissolved in distilled water and stored at room temperature.
3. Running buffer (10X): 25 m*M*, pH 8.3 Tris buffer containing 192 m*M* glycine and 0.1% (w/v) SDS. Store at room temperature.
4. DL-Dithiothreitol solution (DTT) (*see* **Note 1**): 1 *M* of DTT is dissolved in distilled water and stored at −20°C.
5. 2X sample buffer: Laemmli sample buffer (Bio-Rad) and 1 *M* DTT solution are mixed at the ratio of 4:1 (v/v) and stored at −20°C.

6. Resolving gel buffer: 1.5 *M*, pH 8.8 Tris-HCl buffer.
7. Stacking gel buffer: 0.5 *M*, pH 6.8 Tris-HCl buffer.

2.3. Enzyme Purification and Desalting

1. Binding buffer (purification): 20 m*M*, pH 7.4 HEPES buffer with 500 m*M* NaCl and 20 m*M* imidazole (*see* **Note 4**), filtrated by 0.2 μm membrane.
2. Elution buffer (purification): 20 m*M*, pH 7.4 HEPES buffer with 500 m*M* NaCl and 500 m*M* imidazole, filtrated by 0.2 μm membrane.
3. Elution buffer (desalting): 20 m*M*, pH 7.5 HEPES buffer, filtrated by 0.2 μm membrane.

2.4. SWCNT Modification and Enzyme Immobilization

1. Washing buffer: 20 m*M*, pH 7.5 HEPES, filtrated by 0.2 μm membrane.
2. NaOH solution: 2 *M* of NaOH is dissolved in distilled water.
3. Acid mixture: 98% sulfuric acid and 70% nitric acid are mixed at the ratio of 3:1 (v/v).
4. *N*-Hydroxysuccinimide (NHS) solution: 500 m*M* of NHS is dissolved in 20 m*M*, pH 7.5 HEPES buffer.
5. 1-Ethyl-3-[3-(dimethylamino)propyl]carbodiimide (EDC) solution: 100 m*M* of EDC is dissolved in 20 m*M*, pH 7.5 HEPES buffer (*see* **Note 5**).
6. N_α,N_α-bis(carboxymethyl)-L-lysine hydrate (ANTA) solution: 6 m*M* of ANTA is dissolved in 20 m*M*, pH 7.5 HEPES buffer.
7. Super-purified HiPCO® single-walled carbon nanotubes (Unidym).

2.5. Reversible Enzyme Immobilization

1. Elution buffer: 20 m*M*, pH 7.4 HEPES containing 500 m*M* imidazole and 500 m*M* NaCl, filtrated by 0.2 μm membrane.
2. Co^{2+} removing buffer: 20 m*M*, pH 7.4 HEPES with 20 m*M* EDTA, filtrated by 0.2 μm membrane.
3. $CoCl_2$ solution: 100 m*M* $CoCl_2$ in 20 m*M*, pH 7.5 HEPES buffer, filtrated by 0.2 μm filter.
4. Washing buffer: 20 m*M*, pH 7.5 HEPES, filtrated by 0.2 μm membrane.

3. Methods

The effective enzyme immobilization on SWCNTs is based on the specific interaction between the N- or C-terminal His-tag of enzyme and the functionalized SWCNTs. Commercially available

SWCNTs are first treated with acid, ANTA, and $CoCl_2$ to generate the functional SWCNT-ANTA-Co^{2+} complex. Cell lysate containing His-tagged enzyme is then added to the complex to form the SWCNT–enzyme conjugate for catalysis.

It is important to determine efficiency and loading capacity after immobilization. Activity retention can be determined by the same detection method as used in free enzyme characterization. Enzyme-loading capacity can be measured by Bradford assay. To confirm the binding specificity between enzyme and SWCNTs, use samples of cell lysate before and after immobilization and elution solution. To regenerate the SWCNT–enzyme conjugate, "old" enzyme needs to be removed by imidazole first and "new" enzyme can be reloaded by incubating the SWCNTs with freshly prepared cell lysate.

3.1. SWCNTs Modifications and Detections

3.1.1. SWCNT-COOH

1. Weigh 5 mg of super-purified HiPCO® SWCNTs and add to a distillation flask.
2. Add acid mixture (HNO_3:H_2SO_4=1:3 (v/v)) to the distillation flask (**Fig. 8.1**). This treatment can be carried out in a water-bath sonicator at 40°C for 3 h. Adjust pH to pH 3–5 by titrating with 2 *M* NaOH after the reaction.
3. To get the modified SWCNTs, filter the SWCNT dispersion by 0.2 μm membrane and wash with distilled water and 20 m*M*, pH 7.5 HEPES buffer three times.
4. SWCNT-COOH is resuspended in 20 m*M*, pH 7.5 HEPES buffer.

3.1.2. SWCNT-ANTA-Co^{2+}

1. Incubate 60 mM $CoCl_2$ with 6 m*M* ANTA solution overnight. The incubation can be performed on a magnetic stirrer sealed with Para-Film (**Fig. 8.1**).
2. When the incubation finishes, precipitate extra Co^{2+} by adding 2 *M* NaOH to the solution. Remove the blue precipitant $Co(OH)_2$ by centrifugation at 9,000 × *g* for 10 min at room temperature. Collect the supernatant containing ANTA-Co^{2+} in 50 mL centrifuge tubes for further steps.

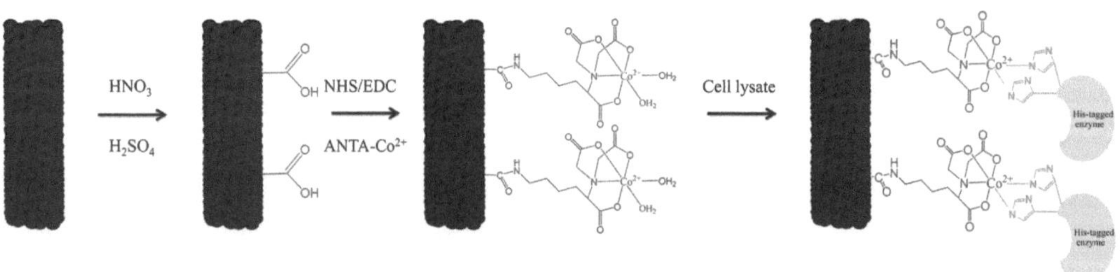

Fig. 8.1. Scheme of SWCNT modification and enzyme immobilization.

3. The SWCNT-ANTA-Co^{2+} complex is produced by overnight amidation of dispersed SWCNT-COOH with ANTA-Co^{2+} in the presence of 100 m*M* NHS, 20 m*M* EDC in 20 m*M*, pH 7.5 HEPES buffer.
4. Collect the SWCNT-ANTA-Co^{2+} complex by filtration through 0.2 μm membrane. Remove excess NHS and EDC by washing SWCNT-ANTA-Co^{2+} complex with 20 m*M*, pH 7.5 HEPES buffer. Suspend the SWCNT-ANTA-Co^{2+} complex in 20 m*M*, pH 7.5 HEPES buffer.

3.1.3. Detection by Fourier Transform Infrared Spectroscopy (FTIR)

1. Prepare potassium bromide (KBr) powder by rubbing its crystal in mortar. Incubate the KBr powder in a 100°C oven to remove water residue.
2. Filter SWCNT samples through 0.2 μm membranes and wash with distilled water three times to remove other impurities.
3. Dry the SWCNT samples in a 100°C oven for 24 h.
4. Mix the dried SWCNT samples with KBr powder and press into thin films for FTIR measurement. The IR spectra are averaged over 128 scans from 1,000 to 4,000 cm^{-1}.

3.2. Enzyme Overexpression and Purification

3.2.1. Enzyme Overexpression

1. Single colony is picked from LB-agar plate and cultured in a 250 mL conical flask containing 30 mL LB medium with appropriate antibiotics at 37°C, 220 rpm overnight.
2. Inoculate 1% (v/v) overnight cell culture to a 1 L conical flask containing 200 mL LB medium and grow cells at 200 rpm, 37°C.
3. Cell growth is monitored by OD at 600 nm using UV-Vis spectrophotometer. When OD of cell culture reaches 0.6–0.8, 0.2–0.4 m*M* IPTG is added into the cell culture to induce overexpression at 37°C for 3 h.
4. After overexpression, cell pellet is collected by centrifugation at 5,000 × *g*, 4°C, 15 min in 50 mL BD falcon tubes and the supernatant is discarded. Cell pellet was washed by 20 m*M*, pH 7.5 HEPES buffer for three times before storing at −20°C.

3.2.2. Enzyme Purification

1. In this process, 10 mL 20 m*M*, pH 7.5 HEPES buffer is added to a falcon tube to suspend cell pellet. The cell suspension is kept in an ice box.
2. Cell lysate is obtained by sonicating the cell suspension in ice bath for 8 × 30 s at an interval of 60 s. Cell lysate and cell debris are separated by centrifugation at 30,000 × *g* at 4°C for 30 min.
3. Commercially available column with either Co^{2+} or Ni^{2+} resin should be equilibrated first with ten-column volume

binding buffer. Cell lysate is added onto the column after equilibrium. Another ten-column volume binding buffer is used to remove nonspecifically bound proteins. Enzyme is eluted with three-column volume elution buffer. After purification, the column is washed with five-column volume elution buffer, ten-column volume DI water, and binding buffer, respectively. Store the column by adding 20% ethanol.

4. PD-10 column (GE healthcare) is first equilibrated with 25 mL 20 m*M*, pH 7.5 HEPES buffer. Purified enzyme solution is loaded onto the column. The enzyme is eluted with 3.5 mL 20 m*M*, pH7.5 HEPES buffer. The column is then washed with 25 mL distilled water and kept in 3.5 mL 20% ethanol at room temperature.
5. SDS-PAGE is used to confirm the enzyme purity. The concentration of enzyme is an average of three-time Bradford measurements using Coomassie protein assay reagent (*see* **Note 6**).

3.3. SDS-PAGE

1. These instructions assume the use of a Bio-Rad Mini-PROTEAN Tetra system. The glass plate for gel casting should be scrubbed clean with suitable detergent every time after use and rinsed with distilled water.
2. Assemble the short plate and spacer plate with cast frame and put it on the cast stand. Make sure that the plate is well cast to prevent the gel leaking.
3. Prepare a 1.0 mm thick 12% resolving gel (**Table 8.1**). Pour the gel into the spacer but leave the space for the stacking gel. Overlay the gel with isobutanol or 0.2% SDS solution and wait for gel polymerization for about 30 min.
4. Remove the overlay solution and wash the top of the resolving gel with DI water.

Table 8.1
12% Resolving gel recipe

Component	Volume (mL)
Distilled water	1.6
30% (29:1) acrylamide/bis solution (*see* **Note 2**)	2.0
1.5 *M* Tris-HCl, pH 8.8	1.3
10% ammonium persulfate solution	0.05
10% SDS solution	0.05
TEMED (*see* **Note 3**)	0.002

Table 8.2
5% Stacking gel recipe

Component	Volume (mL)
Distilled water	1.4
30% (29:1) acrylamide/bis solution (*see* **Note 2**)	0.33
0.5 *M* Tris-HCl, pH 6.8	0.25
10% ammonium persulfate solution	0.02
10% SDS solution	0.02
TEMED (*see* **Note 3**)	0.002

5. Prepare a 1.0 mm thick 5% stacking gel (**Table 8.2**). Quickly pour the stacking gel and insert the comb. The polymerization process takes around 30 min.
6. Prepare loading samples by mixing equal volume of sample buffer and sample. Boil the loading samples for 10 min at 100°C. Centrifuge the samples at 10,000 × *g* for 2 min before loading.
7. Prepare fresh 1X running buffer by diluting 100 mL 10X running buffer with 900 mL distilled water.
8. When the polymerization finishes, wash two glass plates with distilled water and assemble them with the gel running unit. Add running buffer to the upper and lower chambers of the unit. The comb needs to be removed carefully.
9. Load the samples and protein weight marker into wells and connect the gel running unit to the power supply. The initial voltage is 120 V through the stacking gel, and 200 V through the resolving gel.
10. After electrophoresis, get the gel from the glass plates carefully and wash it with 200 mL distilled water for 30 min. Add 100 mL Bio-safe Coomassie Blue and stain the gel for 30 min. The gel needs to be washed and destained by distilled water overnight.

3.4. Enzyme Immobilization

1. Cell lysate is prepared using the same method as in **Section 3.2.2**, Step 2.
2. The SWCNT-ANTA-Co^{2+} complex is dispersed in 20 m*M*, pH 7.5 HEPES buffer by sonication in ice bath.
3. Cell lysate is added to the SWCNT-ANTA-Co^{2+} complex in a beaker sealed with Para-Film. The solution is mixed at 4°C overnight.
4. SWCNT-ANTA-Co^{2+}-enzyme conjugate (SWCNT–enzyme) is collected by centrifuging at 20,100 × *g*, 4°C for

5 min. Supernatant is collected and checked by SDS-PAGE. SWCNT–enzyme is washed with 20 m*M*, pH 7.4 HEPES containing 20 m*M* imidazole for three times to remove nonspecifically bound proteins. The SWCNT–enzyme is then suspended in 20 m*M*, pH 7.5 HEPES buffer.

5. Loading capacity is determined by eluting enzyme off the SWCNT-ANTA-Co^{2+} complex and measuring enzyme concentration with Bradford assay. SWCNT–enzyme is mixed with 20 m*M*, pH 7.5 HEPES containing 500 m*M* NaCl and 500 m*M* imidazole at room temperature overnight. The mixture is separated by centrifuging at 20,100 × *g* for 5 min at room temperature. The SWCNTs left in the supernatant is removed by centrifuging at 100,000 × *g* for 1 h (*see* **Note 7**). The enzyme concentration in the supernatant is measured by Bradford assay.
6. Use samples of cell lysate before and after immobilization and enzyme elution solution to do SDS-PAGE to confirm the binding specificity during immobilization.

3.5. Activity Measurement of Free and Immobilized Enzymes

1. Prepare fresh substrate stock solution and reaction buffer according to the requirements of the enzyme in study. Pay attention to the buffer salt type (*see* **Note 8**), concentration, and pH (*see* **Note 9**).
2. Choose an appropriate detection method to detect enzyme activity. For example, choose UV-Vis spectroscopy if substrate, cofactor, or product has UV absorbance.
3. Set up standard reaction conditions, including buffer pH, reaction volume, and temperature.
4. Add reaction buffer, substrate, cofactor (if necessary), free or immobilized enzyme (add last), mix them together to start the reaction, and monitor the initial reaction rate.
5. Calculate specific activity (μmol min/mg or Unit/mg) of both free and immobilized enzymes. Specific activity is defined as the amount of product formed/substrate consumed per unit time per milligram enzyme under given conditions. Activity retention of immobilized enzyme = (specific activity of immobilized enzyme/specific activity of free enzyme × 100%).

3.6. SWCNT–Enzyme Reversible Immobilization

3.6.1. SWCNT–Enzyme Reloading

1. Enzymes on SWCNTs are eluted off by 20 m*M*, pH 7.4 HEPES containing 500 m*M* NaCl and 500 m*M* imidazole (**Fig. 8.2**). The incubation takes place in an orbital shaker at 200 rpm, room temperature for 1 h.
2. SWCNTs and enzymes are separated by centrifuging at 20,100 × *g* for 5 min at room temperature. The SWCNT-ANTA-Co^{2+} complex is washed with 20 m*M*, pH 7.5

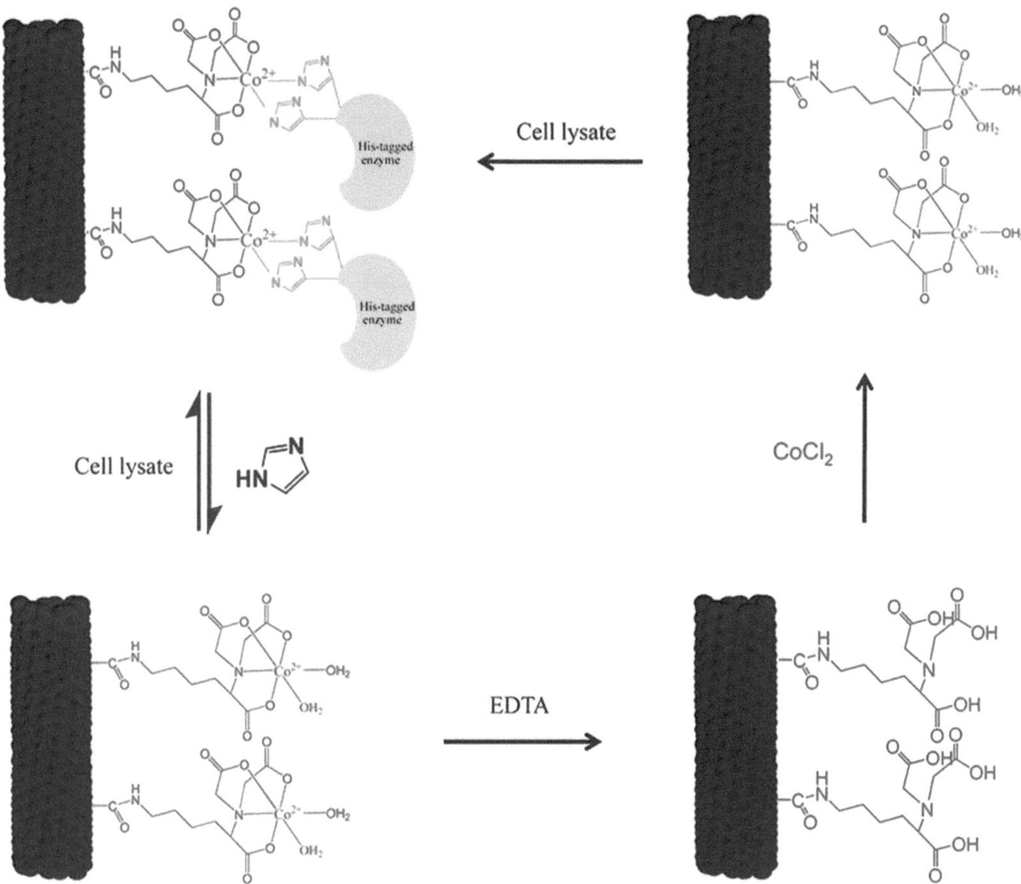

Fig. 8.2. Reversible enzyme immobilization by direct reloading and Co^{2+} regeneration.

HEPES buffer for three to five times to remove imidazole residues.

3. The eluted enzyme concentration is measured by Bradford assay to determine the elution efficiency. The activity of the SWCNT-ANTA-Co^{2+} complex and the eluted enzyme is checked to make sure the removal of enzymes from the SWCNTs.
4. The SWCNT-ANTA-Co^{2+} complex is then dispersed in 20 m*M*, pH 7.5 HEPES by sonication in ice bath.
5. Fresh cell lysate containing overexpressed enzyme is prepared. The enzyme is reloaded onto the dispersed SWCNT-ANTA-Co^{2+} complex and purified as described in **Section 3.4**.
6. The loading capacity and activity retention of the regenerated SWCNT–enzyme are determined by the same method as described previously and compared with that of the original SWCNT–enzyme.

3.6.2. SWCNT–Enzyme Reloading with Co^{2+} Regeneration

1. Enzymes on the SWCNTs are eluted by 20 m*M*, pH 7.4 HEPES buffer with 500 m*M* NaCl and 500 m*M* imidazole at 200 rpm, room temperature for 1 h.
2. SWCNTs and enzyme are separated by centrifuging at 20,100 × *g*, room temperature for 5 min. The SWCNT-ANTA-Co^{2+} complex is washed with 20 m*M*, pH 7.5 HEPES buffer for three to five times to remove imidazole residues.
3. The enzyme concentration of the elution solution is measured by Bradford assay to determine the elution efficiency. The activity of the SWCNT-ANTA-Co^{2+} complex and the elution is also checked to make sure the removal of enzymes from the SWCNTs.
4. Co^{2+} ion from the SWCNT-ANTA-Co^{2+} complex is eluted by 20 m*M*, pH 7.5 HEPES buffer with 500 m*M* NaCl and 20 m*M* EDTA (**Fig. 8.2**). SWCNT-ANTA is washed with 20 m*M*, pH 7.5 HEPES buffer for five times to remove EDTA residues.
5. The SWCNT-ANTA complex is dispersed in 20 m*M*, pH 7.5 HEPES by sonication in ice bath.
6. The SWCNT-ANTA-Co^{2+} complex is regenerated by incubating dispersed SWCNT-ANTA complex in 100 m*M* $CoCl_2$ solution overnight. Excess Co^{2+} ion is removed by washing the SWCNT-ANTA-Co^{2+} complex with 20 m*M*, pH 7.5 HEPES buffer.
7. Enzyme reloading is carried out with fresh cell lysate and dispersed SWCNT-ANTA-Co^{2+} complex using the SWCNT-ANTA complex as control.
8. Enzyme-loading capacity and activity retention of the regenerated SWCNT–enzyme are determined with the same method as stated previously. The original SWCNT–enzyme is selected as control. SDS-PAGE is used to check the reloading efficiency.

4. Notes

1. DTT is a relatively unstable compound and has strong smell. For storage, close bottle tightly and store at 4°C. Handle DTT in a chemical hood.
2. Acrylamide is a light-sensitive and dangerous reagent. Store in an amber bottle at room temperature and wear gloves at any time while handling it.

3. TEMED is a neurotoxic reagent and smells. Store in an amber bottle at room temperature in desiccator. Handle TEMED in a chemical hood with gloves.
4. The imidazole concentration is flexible. Increase its concentration if there are nonspecifically bound proteins found in elution. Decrease the imidazole concentration if the loss of target protein is significant.
5. EDC is harmful by inhalation, in contact with skin, or when swallowed. It is also irritating to eyes and respiratory system. Store it at –20°C in desiccators.
6. Bradford is purchased from Sigma-Aldrich and the assay is conducted according to the manual from supplier.
7. Since SWCNTs can affect the Bradford measurement, they need to be removed from the eluted protein solution before measurement.
8. The choice of salt is determined by two factors: (i) the pK_α value; (ii) if two salts have similar pK_α values, we need to choose the one that does not inhibit the enzyme activity and is comparatively cheaper.
9. Some enzymes are very sensitive to pH changes. The reaction buffer's pH changes when preparation temperature differs from reaction temperature. Therefore, the preparation pH value of reaction buffer needs to be calculated by the Henderson–Hasselbalch equation:

$$\frac{[HA]}{[A^-]}K_\alpha = [H^+].$$

References

1. Wang, P. (2006) Nanoscale biocatalyst systems. *Curr. Opin. Biotechnol.* **17**, 574–579.
2. Asuri, P., Karajanagi, S. S., Dordick, J. S., and Kane, R. S. (2006) Directed assembly of carbon nanotubes at liquid-liquid interfaces: Nanoscale conveyors for interfacial biocatalysis. *J. Am. Chem. Soc.* **128**, 1046–1047.
3. Zhao, X., Jia, H., Kim, J., and Wang, P. (2009) Kinetic limitations of a bioelectrochemical electrode using carbon nanotube-attached glucose oxidase for biofuel cells. *Biotechnol. Bioeng.* **104**, 1068–1074.
4. Kim, M. I., Kim, J., Lee, J., Jia, H., Bin Na, H., Youn, J. K., Kwak, J. H., Dohnalkova, A., Grate, J. W., Wang, P., Hyeon, T., Park, H. G., and Chang, H. N. (2007) Crosslinked enzyme aggregates in hierarchically-ordered mesoporous silica: A simple and effective method for enzyme stabilization. *Biotechnol. Bioeng.* **96**, 210–218.
5. Elgren, T. E., Zadvorny, O. A., Brecht, E., Douglas, T., Zorin, N. A., Maroney, M. J., and Peters, J. W. (2005) Immobilization of active hydrogenases by encapsulation in polymeric porous gels. *Nano Lett.* **5**, 2085–2087.
6. Zhu, Y., Kaskel, S., Shi, J., Wage, T., and van Pee, K. H. (2007) Immobilization of Trametes versicolor laccase on magnetically separable mesoporous silica spheres. *Chem. Mat.* **19**, 6408–6413.
7. Asuri, P., Bale, S. S., Pangule, R. C., Shah, D. A., Kane, R. S., and Dordick, J. S. (2007) Structure, function, and stability of enzymes covalently attached to single-walled carbon nanotubes. *Langmuir* **23**, 12318–12321.

8. Hudson, S., Cooney, J., and Magner, E. (2008) Proteins in mesoporous silicates. *Angew. Chem. Int. Ed.* **47**, 8582–8594.
9. Asuri, P., Karajanagi, S. S., Yang, H. C., Yim, T. J., Kane, R. S., and Dordick, J. S. (2006) Increasing protein stability through control of the nanoscale environment. *Langmuir* **22**, 5833–5836.
10. Yim, T. J., Liu, J. W., Lu, Y., Kane, R. S., and Dordick, J. S. (2005) Highly active and stable DNAzyme–carbon nanotube hybrids. *J. Am. Chem. Soc.* **127**, 12200–12201.

Chapter 9

A TiO_2 Nanoparticle System for Sacrificial Solar H_2 Production Prepared by Rational Combination of a Hydrogenase with a Ruthenium Photosensitizer

Erwin Reisner and Fraser A. Armstrong

Abstract

A hybrid system comprising a hydrogenase and a photosensitizer co-attached to a nanoparticle serves as a rational model for fast dihydrogen (H_2) production using visible light. This chapter describes a stepwise procedure for preparing TiO_2 nanoparticles functionalized with a hydrogenase from *Desulfomicrobium baculatum* (*Db* [NiFeSe]-H) and a tris(bipyridyl)ruthenium photosensitizer (RuP). Upon irradiation with visible light, these particles produce H_2 from neutral water at room temperature in the presence of a sacrificial electron donor – a test-system for the cathodic half reaction of water splitting. In particular, we describe how a hydrogenase and a photosensitizer with desired properties, including strong adsorption on TiO_2, can be selected by electrochemical methods. The catalyst *Db* [NiFeSe]-H is selected for its high H_2 production activity even when H_2 and traces of O_2 are present. Adsorption of *Db* [NiFeSe]-H and RuP on TiO_2 electrodes results in high electrochemical and photocatalytic activities that translate into nanoparticles exhibiting efficient light harvesting, charge separation, and sacrificial H_2 generation.

Key words: Hydrogenase, H_2 production, ruthenium, titanium dioxide, electrochemistry, photochemistry.

1. Introduction

Coupling of a ruthenium dye sensitized TiO_2 nanoparticle to a hydrogenase for sacrificial H_2 production from neutral water based on a previously reported system is described in detail (1, 2). Hydrogenases are metalloenzymes that are expressed in many microorganisms and catalyze the oxidation of H_2 and its production from protons in water, $H^+(aq)$, and electrons.

P. Wang (ed.), *Nanoscale Biocatalysis*, Methods in Molecular Biology 743,
DOI 10.1007/978-1-61779-132-1_9,

Protein film electrochemistry reveals the potential-dependent catalytic properties of these enzymes through the current response observed during substrate turnover (3, 4). Any catalyst suitable for practical applications of photocatalytic H_2 evolution must have fast H_2 production rates, little inactivation by its product (H_2) and other inhibitors (e.g., O_2 and CO), and high stability on the photocatalyst. An unusual [NiFe]-containing hydrogenase having a selenocysteine as one of the ligands to the active site Ni can be isolated from the sulfate-reducing bacterium *Desulfomicrobium baculatum* (*Db*). This hydrogenase, known as *Db* [NiFeSe]-H, shows a degree of O_2 tolerance for H_2 production and relatively weak product inhibition compared to other [NiFe]-hydrogenases, for which H_2 is a strong inhibitor of H_2 production (5). We also discovered that it displays stable attachment and high activity on a TiO_2 electrode, and it was therefore selected for our nanoparticle/enzyme hybrid system.

As photosensitizer we chose a tris(bipyridyl)-ruthenium complex with phosphonic acid linkers [Ru(bipy)$_2$(4,4′-(PO_3H_2)$_2$-bipy)]Br_2 (RuP; bipy = 2,2′-bipyridine). This complex, known as RuP, forms a stable attachment to a TiO_2 surface even at neutral pH. It serves two purposes: (i) it allows visible light harvesting from the solar spectrum, in contrast to using a photosensitizer-free hydrogenase–TiO_2 particle system which requires UV band-gap irradiation ($\lambda < 380$ nm) (6, 7) and (ii) it avoids the formation of reactive (highly oxidizing) electron holes in the valence band of TiO_2, which cause photo-degradation of amino acids adsorbed on TiO_2 (8).

In this methodological chapter, we provide first a stepwise description of the electrochemical studies leading to the selection of *Db* [NiFeSe]-H and RuP for this demonstration. Subsequently, both components are co-attached on a TiO_2 nanoparticle for solar H_2 production in a pH-neutral solution containing a sacrificial electron donor buffer. In this system RuP harvests visible light and injects an electron via the conduction band of TiO_2 into the hydrogenase, resulting in the reduction of protons from water (1, 2). Our rational strategy for the selection of the individual components and assembly of the system is summarized as follows (**Fig. 9.1**):

1. Stable adsorption of *Db* [NiFeSe]-H in an electroactive form on TiO_2 (technique: protein film electrochemistry).
2. Attachment of the photosensitizer RuP to TiO_2 (techniques: electronic absorption spectroscopy and chronoamperometry).
3. Photocatalytic sacrificial H_2 evolution by *Db* [NiFeSe]-H attached on a RuP-sensitized TiO_2 nanoparticle (technique: gas chromatography).

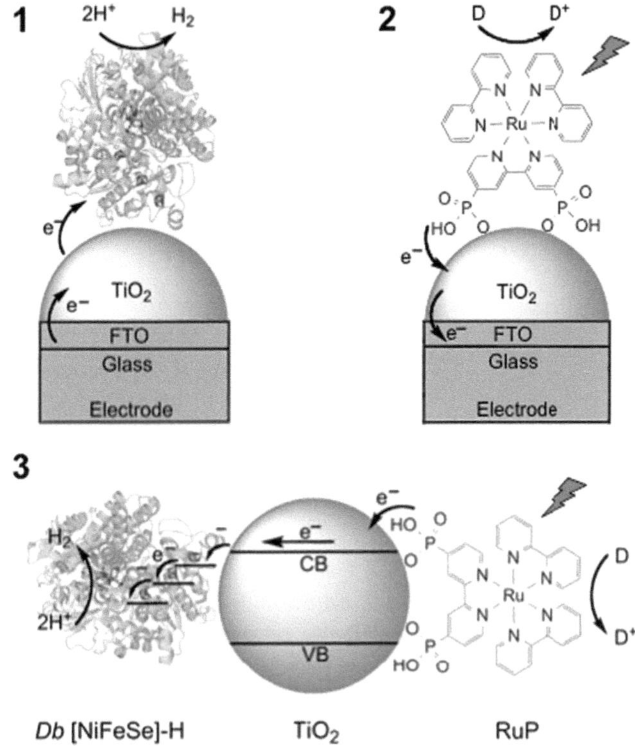

Fig. 9.1. Schematic representation of studies of *Db* [NiFeSe]-H by protein film electrochemistry (**1**; **Section 3.1**), and RuP by photocurrent measurements (**2**; **Section 3.2**) at a nanoparticle TiO_2 electrode, leading to their selection for a demonstration of their suitability for "solar" H_2 production. Visible light driven H_2 evolution with RuP–TiO_2–H_2ase, i.e., *Db* [NiFeSe]-H attached to a RuP dye sensitized TiO_2 nanoparticle (**3**; **Section 3.3**).

2. Materials

1. Starting reagents and materials are of the highest available purity from commercial suppliers, and used as received unless otherwise noted. Purified water (Millipore: 18 MΩ cm) is used. A pH 7.0 triethanolamine buffer (25 mM) is used for all experiments, and for electrochemical experiments 0.10 M NaCl is added as supporting electrolyte. The gases H_2 (Premier Grade, Air Products) and N_2 (oxygen-free, BOC) are introduced in protein film electrochemical experiments by bubbling gently through the cell solutions or passing over the surface. For the photocatalytic experiments the reaction vessel atmosphere (N_2, BOC) contains 2% CH_4 as internal chromatographic standard.
2. Samples of *Db* [NiFeSe]-H were generously provided by the laboratory of Dr. J. C. Fontecilla-Camps (Grenoble,

France). The enzyme was purified from French pressed *D. baculatum* cells using a previously published method (9). *Db* [NiFeSe]-H should be handled anaerobically wherever practical, preferably using a glovebox (Vacuum Atmospheres) under a N_2 atmosphere (O_2 < 2 ppm).

3. Titanium dioxide nanoparticles (Evonik P25-TiO_2, 21 nm particle size; BET: 50 ± 15 m^2/g; 80% anatase, 20% rutile), and indium (ITO) and fluorine-doped SnO_2 (FTO; Hartford Glass TEC-8, 2.3 mm thick) coated glass substrates as rectangular pieces with dimensions 1 × 5 cm were used.
4. Electrochemical experiments are performed using an Autolab PGSTAT 20 electrochemical analyzer controlled by GPES software (Eco Chemie, The Netherlands). A thermostated two-compartment three-electrode glass electrochemical cell is used with a platinum auxiliary electrode and a saturated calomel reference electrode (SCE) held in a sidearm separated from the main cell compartment and linked by a Luggin capillary. Cyclic voltammograms are typically recorded between –0.4 and –0.8 V vs. SCE under a H_2 or N_2 atmosphere, at a scan rate of 10 mV/s at 25°C.
5. Electronic absorption spectroscopy is carried out using a GBC Cintra 10 UV–visible spectrometer. Centrifuged supernatant solutions resulting from the experiments investigating dye adsorption and desorption on TiO_2 are filtered with an Acrodisc 13 mm diameter nylon membrane syringe filter (Aldrich, 0.2 μm) prior to measurements.
6. Photocatalytic experiments are performed with a slide projector (Kodak Carousel S-AV 1010) equipped with a 250 W (24 V) tungsten halogen lamp (Philips). The UV radiation is filtered out with a 420 nm cut-off filter (UQG Optics). The light beam is focused with the projector lenses and light intensity is measured with a Melles Griot Broadband Power/Energy Meter 13PEM001. The reactor for H_2 production consists of a thick-walled pressure reaction vessel (total volume 9 mL) equipped with a stir bar and sealed tightly with a rubber septum. The vessel is thermostated at 25°C with a water circulator connected to a water-jacket reservoir.
7. Detection and quantification of H_2 are achieved using an Agilent 7890A Series gas chromatograph (GC) with electronic pneumatic control optimized for trace H_2 gas analysis. Headspace gas samples (10 μL) from the reaction mixture are injected into the GC using a gas-tight Hamilton syringe in pulsed splitless mode. A 1.5 mm diameter split/splitless liner, a 5 Å molecular sieve column, a micro thermal conductivity detector, and high-purity N_2 carrier gas with a

flow rate of 3.5 mL/min are used. The column oven is held isothermally at 40°C. The reaction vessel is purged with 2% CH_4 (as internal standard for H_2 quantification) in N_2 (2% CH_4/N_2) prior to irradiation. The instrument is calibrated with various known amounts of H_2 in 2% CH_4/N_2 and checked for linearity and stability regularly.

3. Methods

3.1. Protein Film Voltammetry of *Db* [NiFeSe]-H

1. Sonicate a suspension of TiO_2 nanoparticles (50 mg) in 0.10 M nitric acid (80 μL) for 10 min, and spread the viscous dispersion over the doped tin-oxide-coated glass (ITO or FTO) using a Pasteur pipette and a glass slide. An adhesive tape spacer determines the surface area (0.25 cm^2) of the thin transparent film.
2. Dry the TiO_2 film at room temperature for 2 h and then anneal at 450°C for 0.5 h in air. The TiO_2 film is protected from dust and the electrodes are typically used on the day of preparation (*see* **Note 1**).
3. The TiO_2 working electrode (no enzyme) is dipped into the electrochemical cell containing a pH 7 buffer solution thermostated by a water jacket at 25°C (**Fig. 9.2**). A control experiment shows no substantial catalytic H_2 oxidation or

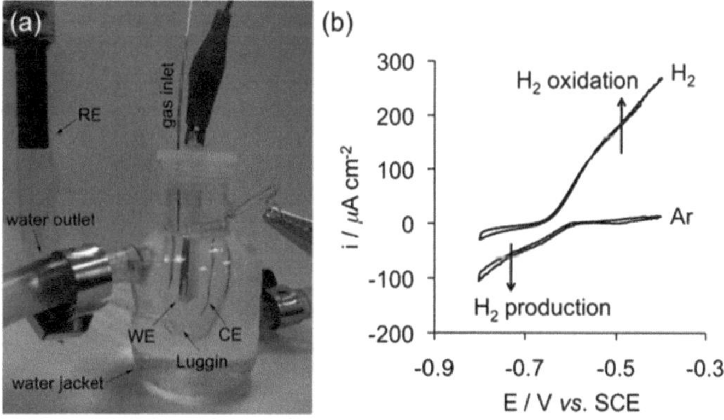

Fig. 9.2. (**a**) Thermostated two-compartment three-electrode glass electrochemical cell used for protein film electrochemistry. The cell features a platinum counter electrode (*CE*) and a saturated calomel reference electrode (*RE*), with the nanoparticle TiO_2 working electrode (*WE*) pointing at the Luggin capillary linking to the RE compartment. The gases H_2 or N_2 are bubbled gently through the solution during measurements to control the composition of the atmosphere through the gas inlet. (**b**) Characteristic protein film voltammogram of *Db* [NiFeSe]-H adsorbed on a stationary TiO_2-FTO electrode with catalytic H_2 production (under Ar) and oxidation (under H_2) at potentials close to the standard reduction potential of the H_2/H^+ redox couple.

proton reduction current by cyclic voltammetry (10 mV/s) between −0.4 and −0.8 V vs. SCE in a H_2 or N_2 atmosphere (*see* **Note 2**).

4. Prepare the *Db* [NiFeSe]-H film by dropping 5 μL of a 1 μM solution of the enzyme (pH 7) onto the TiO_2 surface in an anaerobic glovebox (N_2) and leave adsorbing for 1 min. Rinse carefully with water and dip the electrode into the electrochemical cell with TEOA buffer (25 mM with 0.1 M NaCl).
5. Voltammograms of *Db* [NiFeSe]-H on a TiO_2 working electrode at 25°C under N_2 and H_2 atmospheres show catalytic currents for H_2 production and oxidation, respectively (**Fig. 9.2**, *see* **Note 3**).
6. The decrease in current with time is a measure of the stability of adsorbed enzyme on the electrode surface. *Db* [NiFeSe]-H shows high stability on unmodified TiO_2 and up to 80% of initial current is retained after 2 days when keeping the electrodes in buffered solution at pH 7 (*see* **Note 4**).

3.2. Selection of Ruthenium Photosensitizer

3.2.1. Synthesis of [Ru(bipy)$_2$(4,4′-(PO$_3$Et$_2$)$_2$bipy)]Br$_2$ (bipy = 2,2′-bipyridine), RuP

1. Prepare tetraethyl 2,2′-bipyridine-4,4′-diylbis(phosphonate) by palladium-catalyzed cross-coupling reaction of 4,4′-dibromo-2,2′-bipyridine with diethyl phosphite (10).
2. Replace both chlorido ligands of *cis*-[$Ru^{II}Cl_2$(bipy)$_2$] by tetraethyl 2,2′-bipyridine-4,4′-diylbis(phosphonate), with subsequent hydrolysis of the phosphonate esters to the free phosphonic acids by Me_3SiBr in DMF (**Scheme 9.1**) (11).

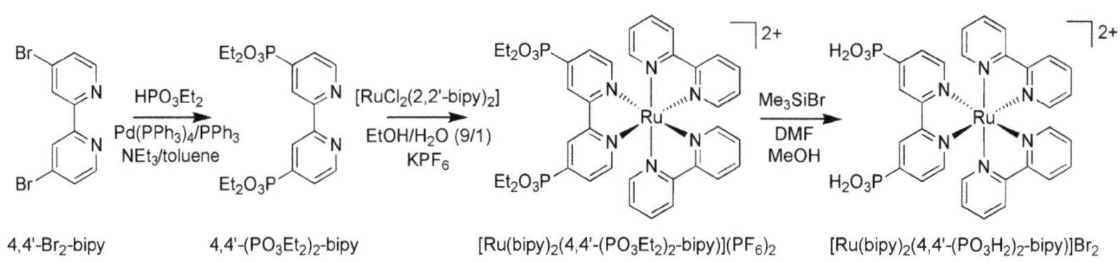

4,4′-Br$_2$-bipy → (HPO_3Et_2; Pd(PPh$_3$)$_4$/PPh$_3$; NEt$_3$/toluene) → 4,4′-(PO$_3$Et$_2$)$_2$-bipy → ([RuCl$_2$(2,2′-bipy)$_2$]; EtOH/H$_2$O (9/1); KPF$_6$) → [Ru(bipy)$_2$(4,4′-(PO$_3$Et$_2$)$_2$-bipy)](PF$_6$)$_2$ → (Me$_3$SiBr; DMF; MeOH) → [Ru(bipy)$_2$(4,4′-(PO$_3$H$_2$)$_2$-bipy)]Br$_2$

Scheme 9.1. Synthetic pathway to RuP.

3.2.2. Adsorption and Stability of RuP on TiO_2

1. Sonicate TiO_2 (5 mg) in 1.5 mL water for 5 min, add (dropwise) an aqueous unbuffered 0.20 mM RuP solution (0.5 mL) to the stirred TiO_2 dispersion, and continue stirring the yellow-orange mixture overnight, protected from light.
2. Separate the particles from the supernatant by centrifugation and dry under reduced pressure. Confirm quantitative adsorption by the disappearance of the d → π^* band (λ_{max} = 456 nm) in the colorless supernatant of the centrifuged RuP-sensitized TiO_2 dispersion (*see* **Note 5**).

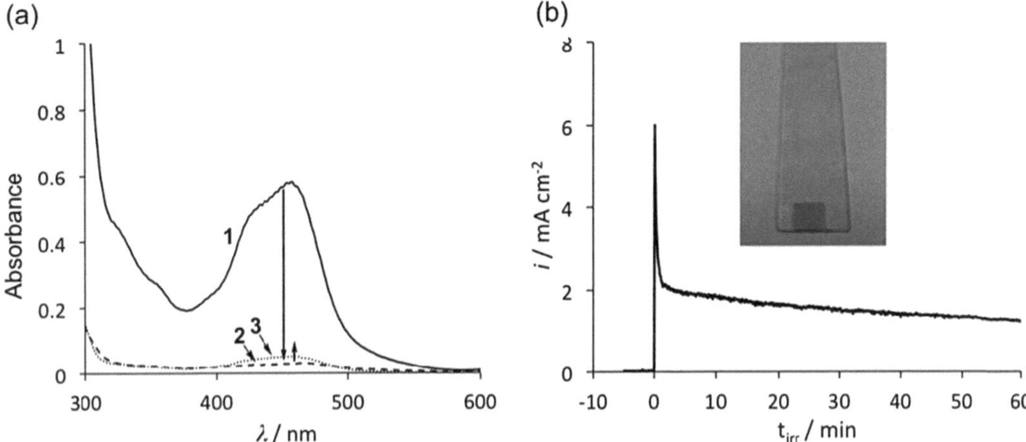

Fig. 9.3. (**a**) Electronic absorption spectrum of an aqueous solution of RuP (0.05 mM in 25 mM triethanolamine, pH 7) before (*1*) and after (*2*) adsorption on TiO_2 (5 mg TiO_2/2 mL 0.05 mM RuP solution). *3*: The UV–vis spectrum of the supernatant solution after stirring RuP-sensitized TiO_2 for 4 h in a pH 7 triethanolamine buffer (25 mM, 2 mL): the spectrum shows minimal desorption of RuP from TiO_2. (**b**) Visible light driven ($\lambda > 420$ nm) photocurrent obtained with a RuP-sensitized TiO_2-FTO electrode in pH 7.0 triethanolamine buffer with an applied voltage of 0.5 V vs. SCE in air. The first few minutes were a dark period (no current) followed by visible light illumination at $t_{irr} = 0$. The *insert* shows a typical RuP-sensitized TiO_2-FTO electrode.

3. Test the stability of RuP–TiO_2 by dispersing the particles with ultrasound treatment in pH 7 buffered TEOA solution (2 mL, 25 mM), then stir the mixture for 4 h at room temperature.
4. Examine the electronic absorption spectrum of the filtered colorless supernatant, which should show only marginal desorption of dye from TiO_2 (**Fig. 9.3**, *see* **Note 6**).

3.2.3. Photocurrent with RuP-Sensitized TiO_2 Electrodes

1. Prepare photoanode by immersing a freshly annealed nanoparticle TiO_2 electrode (0.25 cm^2 TiO_2 particulate film; prepared as described above) into an aqueous 0.1 mM solution (10 mL) of RuP and leaving to stand overnight.
2. An intense orange coloration of the TiO_2 surface indicates strong attachment of RuP to TiO_2. Remove the RuP-sensitized TiO_2 electrode from the RuP solution, rinse carefully with water, and place it into the electrochemical cell solution containing TEOA (pH 7.0, 25 mM) and 0.1 M NaCl thermostated at 25°C. The electrode is left to equilibrate, undisturbed, for 15 min before proceeding.
3. Photocurrent measurements are performed chronoamperometrically by applying a voltage of 0.5 V vs. SCE in air at 25°C to the RuP-sensitized TiO_2 electrode while irradiating with a focused commercial 250 W tungsten halogen slide projector lamp equipped with a 420 nm UV cut-off filter (*see* **Note 7**).

4. The larger the photocurrent the more electrons are injected from the adsorbed sensitizers into the conduction band of the TiO_2 electrode. The phosphonate-linked ruthenium photosensitizer shows strong chemical attachment to Ti^{IV} even at pH 7. The photocurrent response upon visible light irradiation of RuP-sensitized TiO_2 electrodes (**Fig. 9.3b**) shows efficient direct electron transfer between the ruthenium sensitizer and the semiconductor.

3.3. Photocatalytic H_2 Production with Enzyme–TiO_2 Hybrid System

1. Sonicate TiO_2 nanoparticles (50 mg) in water (4.75 mL) for 5 min, add a solution of RuP (0.9 mg, 1.0 μmol) in water (0.25 mL) dropwise under stirring, and stir the resulting dispersion at room temperature overnight, protected from light by aluminum foil. This procedure results in quantitative adsorption of RuP on TiO_2, i.e., formation of RuP–TiO_2 nanoparticles.
2. The RuP–TiO_2 particles are separated from the colorless aqueous solution by centrifugation, dried under reduced pressure, and separated into 5 mg aliquots (each containing 0.1 μmol of RuP) for the photocatalysis experiments.
3. The RuP–TiO_2 particles (5 mg) are sonicated in pH 7.0 triethanolamine buffer (4.5 mL, 25 mM) in a Pyrex pressure reaction vessel (total volume 9 mL) for 5 min to form a fine dispersion. This operation is carried out in an anaerobic glovebox. Then, *Db* [NiFeSe]-H (20 μL of 1 μM solution) is added dropwise to the stirred dispersion of RuP–TiO_2. The vessel is stirred for 10 min, sealed tightly with a rubber septum, taken out of the glovebox, and purged with 2% CH_4 (as internal chromatographic standard) in N_2 for 10 min (*see* **Note 8**).
4. Photocatalytic experiments are performed by visible light illumination with a focused tungsten halogen projector lamp (45 mW/cm^2) with a UV cut-off filter ($\lambda > 420$ nm) of the stirred and thermostated colloidal particle reaction mixture at 25°C (*see* **Note 9**).
5. Quantification of H_2 in the headspace of the reaction mixture (10 μL) is performed by gas chromatography. In a typical experiment, 6.7% H_2 accumulates after 4 h of irradiation (**Fig. 9.4**). The pH of the reaction mixture after illumination remains between 7.0 and 7.2.
6. Control experiments performed in the absence of RuP, TiO_2 or *Db* [NiFeSe]-H under the same experimental conditions show only negligible or no traces of H_2.

3.4. Conclusion

The attachment of *Db* [NiFeSe]-H and RuP in an electroactive form on a TiO_2 nanoparticle, as studied by protein film

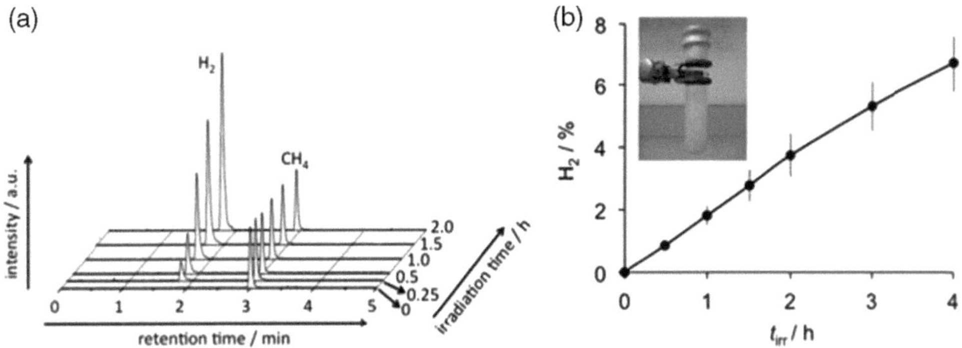

Fig. 9.4. Typical GC traces (**a**) and quantification (**b**) of photocatalytic H_2 production with *Db* [NiFeSe]-H adsorbed on RuP-sensitized TiO_2 at 25°C and in pH 7.0 triethanolamine buffer (25 mM). The *insert* shows an assembled RuP–TiO_2–NiFeSe system.

electrochemistry and photocurrent measurements, is exploited to yield excellent photocatalytic H_2 production activity in an integrated system. Stable attachment of the active enzyme on TiO_2 together with the excellent H_2 production activity and some O_2 tolerance of *Db* [NiFeSe]-hydrogenase are prerequisites for efficient visible light driven H_2 production without rigorous anaerobic conditions.

4. Notes

1. We encountered problems with cracking of nanoparticle TiO_2 films with large (>0.25 cm^2) or thick (non-transparent) surfaces.
2. Particulate TiO_2 film electrodes show characteristic charging/discharging currents to a variable extent at ca. –0.5 to –0.8 V vs. SCE. The background signal from these charging currents can be decreased by using a slower scan rate (e.g., 1 mV/s). All reduction potentials in this chapter are quoted vs. SCE (experimentally obtained values) and can be converted to the standard hydrogen electrode (SHE) by adding +245 mV.
3. A control experiment in which the solution of *Db* [NiFeSe]-H is applied directly on the blank doped tin-oxide surface (in the absence of TiO_2) shows negligible electrocatalytic response (on ITO) or relatively small current densities (on FTO). It should be noted that *Db* [NiFeSe]-H is so far the only hydrogenase for which significant electrocatalytic activity on a TiO_2 electrode has been observed by protein film electrochemistry.

4. The mesoporous TiO_2 film with many pores and channels might be responsible for the high current densities and stability of the enzyme film. The isoelectric point (Ip) of TiO_2 is ca. 6, and the surface charge of the nanoparticles and the [NiFeSe]-hydrogenase (Ip = 5.4) could be important for efficient adsorption. The temporal decrease in current density can have several causes: traditional *film loss*, which refers to a decrease in current due to enzyme desorption, enzyme denaturation and an unfavorable conformational rearrangement on the electrode surface are possible contributing factors. In the case of TiO_2 electrodes, another factor could be TiO_2 *particle loss* from the annealed and plain doped tin-oxide electrodes in buffered aqueous solutions. Particle loss tends to be rather irreproducible (and more pronounced for ITO than FTO), and for the determination of film stability an electrode with negligible particle loss was used.
5. The colorless supernatant is filtered with a 0.2 μm nylon membrane filter prior to UV–vis spectrometry to remove remaining traces of nanoparticles, which cause light scattering and a baseline offset in the UV–vis spectrum.
6. Carboxylate-linked bipyridyl-based ruthenium complexes adsorbed on TiO_2, as used in dye-sensitized solar cells, desorb almost quantitatively from TiO_2 under the experimental conditions employed (buffered neutral aqueous medium), and are not suitable for this system.
7. The lamp should be ignited at least 5 min before illuminating samples to obtain a stable light intensity. The focused light beam is covered prior to irradiation.
8. Triethanolamine buffer is a considerably better sacrificial electron donor in this system than ethylenediamine tetraacetic acid, which causes photobleaching of the photosensitizer.
9. Gentle stirring is very important for long-term stability of the dispersion. Vigorous stirring overnight results in almost complete inactivation of the system (presumably due to accelerated film loss).

Acknowledgments

This work was supported by BBSRC (BB/D52222X/1 and BB/H003878-1) and EPSRC (Supergen 5 and EP/H00338X/1).

References

1. Reisner, E., Fontecilla-Camps, J. C., and Armstrong, F. A. (2009) Catalytic electrochemistry of a [NiFeSe]-hydrogenase on TiO_2 and demonstration of its suitability for visible-light driven H_2 production. *Chem. Commun.* 550–552.
2. Reisner, E., Powell, D. J., Cavazza, C., Fontecilla-Camps, J. C., and Armstrong, F. A. (2009) Visible light-driven H_2 production by hydrogenases attached to dye-sensitized TiO_2 nanoparticles. *J. Am. Chem. Soc.* **131**, 18457–18466.
3. Vincent, K. A., Parkin, A., and Armstrong, F. A. (2007) Investigating and exploiting the electrocatalytic properties of hydrogenases. *Chem. Rev.* **107**, 4366–4413.
4. Armstrong, F. A., Belsey, N. A., Cracknell, J. A., Goldet, G., Parkin, A., Reisner, E., Vincent, K. A., and Wait, A. F. (2009) Dynamic electrochemical investigations of hydrogen oxidation and production by enzymes and implications for future technology. *Chem. Soc. Rev.* **38**, 36–51.
5. Parkin, A., Goldet, G., Cavazza, C., Fontecilla-Camps, J. C., and Armstrong, F. A. (2008) The difference a Se makes? Oxygen-tolerant hydrogen production by the [NiFeSe]-hydrogenase from Desulfomicrobium baculatum. *J. Am. Chem. Soc.* **130**, 13410–13416.
6. Cuendet, P., Rao, K. K., Grätzel, M., and Hall, D. O. (1986) Light induced hydrogen evolution in a hydrogenase-titanium dioxide particle system by direct electron transfer or via rhodium complexes. *Biochimie* **68**, 217–221.
7. Nikandrov, V. V., Shlyk, M. A., Zorin, N. A., Gogotov, I. N., and Krasnovsky, A. A. (1988) Efficient photoinduced electron transfer from inorganic semiconductor titanium dioxide to bacterial hydrogenase. *FEBS Lett.* **234**, 111–114.
8. Hidaka, H., Shimura, T., Ajisaka, K., Horikoshi, S., Zhao, J., and Serpone, N. (1997) Photoelectrochemical decomposition of amino acids on a TiO_2/OTE particulate film electrode. *J. Photochem. Photobiol. A* **109**, 165–170.
9. Hatchikian, E. C., Bruschi, M., Le Gall, J., Forget, N., and Bovier-Lapierre, G. (1978) Characterization of the periplasmic hydrogenase from Desulfovibrio gigas. *Biochem. Biophys. Res. Commun.* **82**, 451–461.
10. Penicaud, V., Odobel, F., and Bujoli, B. (1998) Facile and efficient syntheses of 2,2′-bipyridine-based bis(phosphonic) acids. *Tetrahedron Lett.* **39**, 3689–3692.
11. Trammell, S. A., Moss, J. A., Yang, J. C., Nakhle, B. M., Slate, C. A., Odobel, F., Sykora, M., Erickson, B. W., and Meyer, T. J. (1999) Sensitization of TiO_2 by phosphonate-derivatized proline assemblies. *Inorg. Chem.* **38**, 3665–3669.

Chapter 10

Preparation and Characterization of Single-Enzyme Nanogels

Jun Ge, Ming Yan, Diannan Lu, Zhixia Liu, and Zheng Liu

Abstract

Enzymes have been incorporated in nanostructures in order to provide robust catalysts for valuable reactions, particularly those performed under harsh and denaturing conditions. This chapter describes the encapsulation of enzymes in polyacrylamide nanogels by a two-step in situ polymerization process for preparing robust biocatalysts. The first step in this process is the generation of vinyl groups on the enzyme surface, while the second step involves in situ polymerization using acrylamide as the monomer. Encapsulation of the enzyme in the hydrophilic, porous, and flexible polyacrylamide gel of several nanometers thick would help to both give a significantly enhanced thermostability and prevent the removal of essential water by polar solvents. The hydrophilic flexible polymer shell also allows the protein structure to undergo necessary conformational transitions during the catalytic reaction and, at the same time, impose marginal mass transfer restrictions for the substrates entering across the polymer shell. The effectiveness of this method is demonstrated with horseradish peroxidase (HRP), carbonic anhydrase, and lipase. The impacts of such an encapsulation on the activity and stability of enzymes are also discussed.

Key words: Nanostructured biocatalyst, polyacrylamide nanogel, enzyme encapsulation, enzyme stability, enzymatic catalysis, molecular simulation.

1. Introduction

Recent advances in enzyme chemistry, material science, and computational technology have promoted the study of chemical modification of enzymes to obtain enhanced stability and catalytic performance against adverse conditions that often lead to enzyme deactivation. Incorporating enzymes into polymeric nanostructures (1, 2) is particularly promising because the flexibility in designing structures of the polymers offers immense possibilities

P. Wang (ed.), *Nanoscale Biocatalysis*, Methods in Molecular Biology 743,
DOI 10.1007/978-1-61779-132-1_10, © Springer Science+Business Media, LLC 2011

for tailoring enzymes for various applications such as nonaqueous catalysis (3, 4), bionanodevices (5–8), intelligent molecular machines (9, 10), and artificial cells (11, 12). While some enzymes are stable in hydrophobic organic solvents (13–15), for those hydrophilic organic ones such as methanol, dimethyl sulfoxide (DMSO), and dimethylformamide (DMF), a threshold concentration exists, above which the dissolved enzyme is fully deactivated (16, 17). This is mainly because that these hydrophilic solvents can extract the "essential water" from the vicinity of enzyme surface, leading to the unexpected alternation of enzyme conformation that determines its functionality. Moreover, DMSO and DMF, which are known as universal solvents, were reported to unfold the tertiary structure of enzyme (18). The stabilization of enzyme in hydrophilic organic solvent is thus a challenging objective to enzyme modification. Recently, Liu and coworkers developed an in situ polymerization procedure that produced single-enzyme nanogels with greatly enhanced thermal stability and tolerance against hydrophilic organic solvents (19, 20). The preparation of an enzyme nanogel begins with the generation of vinyl groups on the enzyme surface by acryloylation (21). This is followed by in situ polymerization of acrylamides that eventually leads to encapsulation of the acryloylated enzyme. The efficiency and the versatility of this method have been demonstrated using horseradish peroxidase (HRP) (19), carbonic anhydrase (22), and a lipase (23) as model enzymes. In all cases, the enzyme nanogels had shown essentially the same catalytic activities and kinetics as their native counterparts. However, the enzyme nanogels exhibited substantially higher tolerance against high temperatures and organic solvents than did their native counterparts, which were rapidly deactivated. When applied to the synthesis of a dextran-based surfactant at 60°C in anhydrous DMSO, the *Candida rugosa* lipase nanogel maintained its original activity for 240 h, whereas the native enzyme was completely deactivated within 30 min. Molecular structures of nanogel enzymes have been studied through complementary inputs from molecular dynamics (MD) simulations at all-atom level and various types of structural characterization techniques. It appeared that hydrogen bonding is the major driving force for the formation of the monomer assembly around the enzyme, which is a prerequisite in the preparation of enzyme nanogels (23). In addition to strengthening the structural integrity against thermal fluctuation, the polyacrylamide shell also prevents the loss of surface water from the enzyme molecules into the polar solvents, thus stabilizing the enzymes in anhydrous polar solvents. The porous and flexible nature of the polyacrylamide network, with a thickness of several nanometers, only slightly restricts both the accessibility of the enzyme and its structural transitions during catalysis. This also explains the free enzyme-like catalytic performances of the enzyme nanogels.

It can be expected that the enzyme nanogel method to be described in this chapter will become a useful tool in developing chemically engineered enzymes for production of robust biocatalysts for various applications.

2. Materials

2.1. Fabrication of Single-Enzyme Nanogels

1. Enzyme solution I: Horseradish peroxidase (HRP) dissolved in 100 mM boric acid buffer (pH 9.3) with final protein concentration of 2.5 mg/mL together with 0.25 mg/mL of 4-dimethylaminoantipyrine, which is a reagent that protects HRP against modification (*see* **Note 1**).
2. Enzyme solution II: Bovine carbonic anhydrase II (BCA) dissolved in 100 mM boric acid buffer (pH 9.3) with a protein concentration of 2.5 mg/mL.
3. Enzyme solution III: Lipase from *C. rugosa* (type VII) (CRL) dissolved in 50 mM acetic buffer (pH 4.0) and centrifuged at 10,000 rpm and 4°C for 10 min to remove precipitates (*see* **Note 2**), followed by an adjustment of protein concentration to 2 mg/mL.
4. Acryloylation reagent: *N*-Acryloxysuccinimide (NAS) dissolved in dimethyl sulfoxide (DMSO) to a final concentration of 20 mg/mL prepared prior to use (*see* **Note 3**).
5. Monomers: Monomers and initiators for polymerization, including acrylamide (AM), *N,N′*-methylenebisacrylamide (MBA), *N,N,N′,N′*-tetramethylethylenediamine (TEMED), and ammonium persulfate (APS), used without further purification.

2.2. Structural Characterization of Single-Enzyme Nanogels

All samples of the free enzyme and enzyme nanogel are dialyzed against 10 mM sodium phosphate buffer (pH 7.5) for 12 h at 4°C prior to structural characterization:

1. Size exclusion chromatography (SEC): SEC is performed using a TSK-GEL SW4000×L column coupled to a fluorescence detector. The elution buffer is 0.1 M sodium phosphate (pH 6.7) containing 0.1 M Na_2SO_4 (*see* **Note 4**). The free enzyme and enzyme nanogel samples (protein content, 0.1 mg/mL) are dissolved in 10 mM sodium phosphate buffer (pH 7.5).
2. Transmission electron microscopy (TEM): Enzyme nanogel samples (protein content, 0.1 mg/mL) are dissolved in 10 mM sodium phosphate buffer (pH 7.5). The negative dying reagent sodium phosphotungstate is dissolved

in deionized water to a concentration of 1–2% (w/v) and adjusted to pH 7.0 with 1 M NaOH. It is prepared 2 h prior to staining.

3. Atomic force microscopy (AFM): Enzyme nanogel samples (protein content, 3 μM) are dissolved in 10 mM sodium phosphate buffer (pH 7.5).
4. Scanning electron microscopy (SEM): Enzyme nanogel samples (protein content, 0.1 mg/mL) are dissolved in 10 mM sodium phosphate buffer (pH 7.5).
5. Dynamic light scattering (DLS): Enzyme nanogel samples (protein content, 0.1 mg/mL) are dissolved in 10 mM sodium phosphate buffer (pH 7.5) and filtered through 0.22 μm membrane before measurement.
6. Circular dichroism (CD): The free enzyme and enzyme nanogel samples (protein content, 5–7 μM) are dissolved in 10 mM sodium phosphate buffer (pH 7.5).
7. Fluorescence microscopy (FL): The free enzyme and enzyme nanogel samples (protein content, 0.1 mg/mL) are dissolved in 10 mM sodium phosphate buffer (pH 7.5).
8. Differential scanning calorimetry (DSC): The free enzyme and enzyme nanogel samples (protein content, 5 μM) are dissolved in 10 mM sodium phosphate buffer (pH 7.5).

2.3. Enzymatic Activity Assay

1. Substrate solution for measuring horseradish peroxidase (HRP) activity: Solutions of 1.1 mM H_2O_2 in 100 mM phosphate citrate buffer (pH 5.5), 0.02 M H_2O_2 in water, and 0.02 M 3,3′,5,5′-tetramethylbenzidine (TMB) in DMSO prepared prior to use.
2. Substrate solution for measuring bovine carbonic anhydrase (BCA) activity: A stock solution of *p*-nitrophenyl acetate in acetonitrile.
3. Substrate solution for measuring *C. rugosa* lipase (CRL) activity: *p*-Nitrophenyl palmitate (*p*-NPP) firstly dissolved (3.8 mg) in 1 mL acetone and then diluted with 20 mL sodium phosphate buffer (50 mM, pH 7.0) containing 1.25% (w/v) Triton X-100 (*see* **Note 5**).

3. Methods

3.1. Preparation of Single-Enzyme Nanogels

1. Acryloylation: *N*-acryloxysuccinimide (NAS) dissolved in DMSO is slowly added under gentle stirring at 4°C to 20 mL of the enzyme solution, resulting in NAS:enzyme

molar ratios ranging from 20:1 to 200:1 (*see* **Note 6**). The reaction is continued for 12 h at 4°C.

2. After the acryloylation step, the enzyme solution is collected and dialyzed against 50 mM sodium phosphate buffer (pH 7.0) at 4°C for 24 h to remove the unreacted reagents.
3. Acrylamide in a concentration range of 10–50 mg/mL and *N,N′*-methylenebisacrylamide in a concentration range of 1–5 mg/mL are directly added to 20 mL of the acryloylated enzyme solution (*see* **Note 7**). This is followed by purging with N_2 for 10 min. Subsequently, 60 mg APS and 50 μL TEMED are added to initiate polymerization. The polymerization reaction continues for 6 h at 30°C or 12 h at 4°C.
4. After polymerization, the solution is collected and dialyzed against 50 mM sodium phosphate buffer (pH 7.0) at 4°C for 24 h to remove the unreacted reagents. This is followed by lyophilization for 48 h to obtain the enzyme nanogel in the powder form.

3.2. Characterization of Single-Enzyme Nanogels

Single-enzyme nanogels are characterized by the following three categories: determination of the encapsulation ratio by SEC; determination of the size and morphology by TEM, AFM, scanning SEM, and DLS; and characterization of the secondary and tertiary structures of the encapsulated enzyme by CD, FL, and DSC:

1. SEC with fluorescence detection is used to determine the encapsulation ratio of the enzyme in nanogels. Native enzyme, enzyme nanogel, or the mixture of native enzyme and free polyacrylamide that served as the control (protein content, 0.1 mg/mL) is filtered through a 0.45-μm membrane and subjected to SEC. The elution volume of the encapsulated enzyme is smaller than that of its native counterpart. The encapsulation ratio of the enzyme is determined from the elution peak area of the nanogel with respect to the overall elution areas of both the encapsulated enzyme and the residual enzyme.
2. In the case of TEM, a drop (~10 μL) of an aqueous solution of the enzyme nanogel (protein content, 0.1 mg/mL) is first placed on the carbon-coated grid, and after 30–60 s, the excess is removed with a filter paper. After drying at room temperature for approximately 60 s, a drop (~10 μL) of 2% (w/v) sodium phosphotungstate (pH 7.0) is placed on the carbon-coated grid. After 30–60 s, the excess is removed with a filter paper. The sample is thoroughly dried at room temperature before visualization by TEM. An example of TEM of the HRP nanogel is shown in **Fig. 10.1a** (*see* **Note 8**).

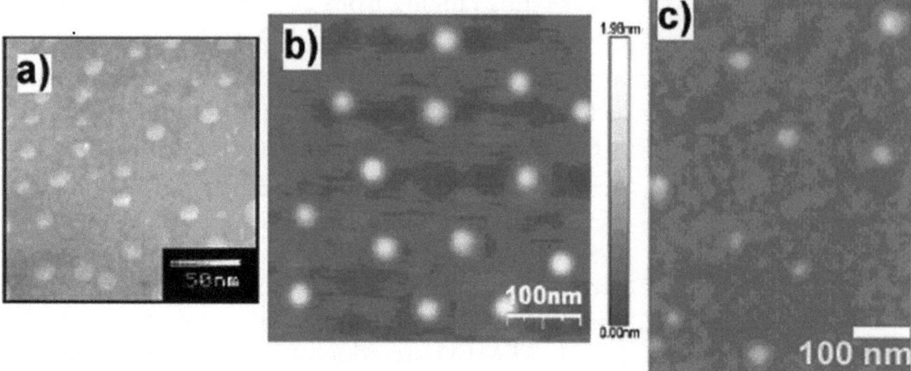

Fig. 10.1. (**a**) TEM of a single HRP nanogel [reproduced from (19) with permission from American Chemical Society (13)]. (**b**) AFM of a single BCA nanogel [reproduced from (22) with permission from American Chemical Society (16)]. (**c**) SEM of a single CRL nanogel [reproduced from (20) with permission from American Chemical Society (20)].

3. In the case of AFM, a new mica surface is prepared by fresh cleaving. It is exposed to a 3 μM enzyme nanogel solution for 30 s, followed by measurements in the tapping mode in air. An example of AFM of a BCA nanogel is shown in **Fig. 10.1b** (*see* **Note 9**).
4. In the case of SEM, a drop (~10 μL) of enzyme nanogel (protein content, 0.1 mg/mL) is placed on the glass surface (*see* **Note 10**), and the excess is removed with a filter paper. The sample is dried at room temperature for 10 days and directly visualized by SEM. An example of SEM of a CRL nanogel is shown in **Fig. 10.1c**.
5. DLS, which is used to determine the size and polydispersity index of the enzyme nanogel, is performed in a DynaPro-801 dynamic light scattering instrument. Data are collected and analyzed using the AutoPro data software for the DynaPro-801 instrument.
6. CD spectroscopy in the range of 180–250 nm is carried out for both the free enzyme and the enzyme nanogel (protein content, 5–7 μM). Each sample is scanned three times at a speed of 20 nm/min with a step length of 0.5 nm.
7. The FL emission spectra of the free enzyme and the enzyme nanogel (protein content, 0.1 mg/mL) are recorded in the range from 285 to 600 nm with an excitation wavelength of 285 nm. This permits characterization of the tertiary structure of the encapsulated enzyme.
8. DSC of the free enzyme and the enzyme nanogel (protein content, 5 μM) is performed to determine the melting temperature of the encapsulated enzyme. Data are recorded when the temperature increases from 30 to 90°C at a rate of 1°C/min.

3.3. Enzymatic Activity Assay

1. Activity measurement for HRP: A 0.9 mL aliquot of phosphate citrate (pH 5.5, 100 mM) containing 1.1 mM H_2O_2, 0.05 mL 0.02 M H_2O_2, and 10 μL 0.2 μg/mL HRP is mixed, and 0.05 mL DMSO containing 0.02 M TMB is added to initiate the reaction. The oxidation rate of TMB is calculated from the slope of the initial linear part of the adsorption curve at 655 nm using a molar absorption coefficient of 39,000 M/cm for the oxidation product of TMB (24).
2. Activity measurement for BCA: The esterase activity of BCA is determined according to the method described by Pocker and Stone (25, 26). A stock solution of *p*-nitrophenyl acetate is dissolved in acetonitrile. The initial rate of product formation is determined at 348 nm using a molar absorption coefficient of 5,000 M/cm. A blank solution is used as the control.
3. Activity measurement for CRL: *p*-NPP is first dissolved in acetone and then diluted in sodium phosphate buffer (50 mM, pH 7.0) containing 1.25% (w/v) Triton X-100 to obtain a final concentration of 0.5 mM. The reaction is initiated by adding 50 μL enzyme solution to 950 μL of the above substrate solution, and detection is carried out at 348 nm (27).

3.4. Stability of the Native Enzyme and Enzyme Nanogel at High Temperatures and in the Presence of Organic Solvents

To determine the thermal stability of the enzyme nanogel, the native enzyme and the enzyme nanogel, which had identical protein contents, are incubated at a high temperature for a given time. The preparations are sampled at specific time intervals, and their enzymatic activity was measured. To determine the stability in organic solvents, the native enzyme and the enzyme nanogel, which had identical protein contents, are first incubated in a solution with a given content of organic solvent for a specified time. Samples are removed at specific time intervals, and their enzymatic activity is measured. For example, **Fig. 10.2a** shows that the HRP nanogel maintained 80% of its initial activity after 90 min incubation at 65°C, while free HRP lost all its activity after incubation for only 30 min. **Figure 10.2b** shows that at 75°C, the half-life of BCA increased from several minutes to 100 min after encapsulation in a nanogel of diameter 18.2 nm. **Figure 10.2c** shows that the activity of the CRL nanogel was almost unchanged over a period of 500 min at 50°C. This is in contrast to the behavior of the native lipase that lost 50% of its original activity within 30 min. **Figure 10.2d** shows that after 3 h incubation in anhydrous DMSO or DMF at 50°C, native CRL is completely deactivated, while the CRL nanogel retained 80 and 120% of its original activity, respectively.

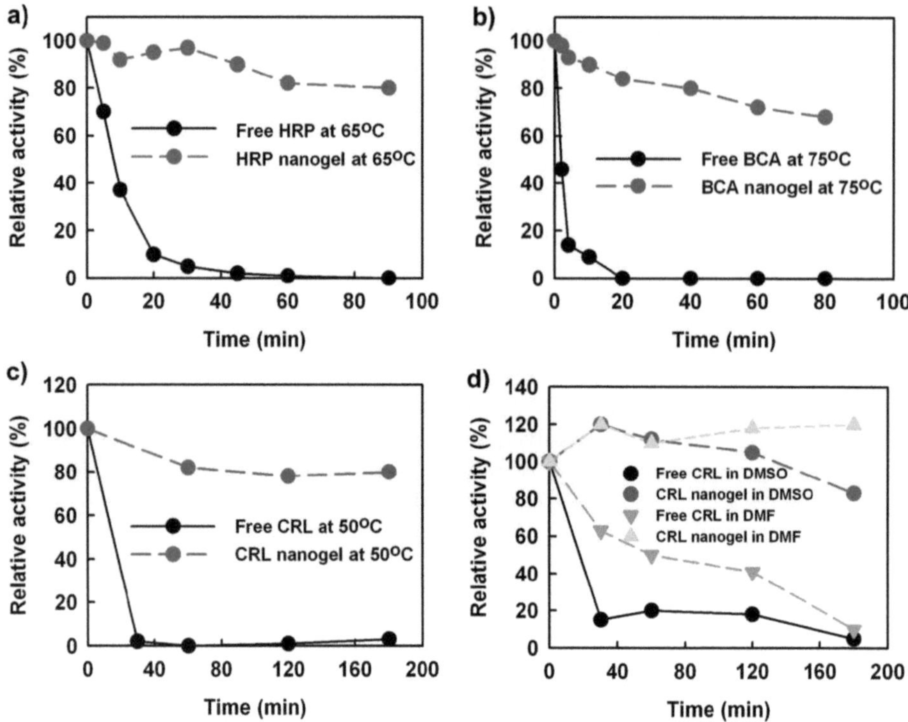

Fig. 10.2. (**a**) Stability of free HRP and the HRP nanogel at 65°C. (**b**) Stability of free BCA and the BCA nanogel at 75°C. (**c**) Stability of free CRL and the CRL nanogel at 50°C. (**d**) Stability of free CRL and the CRL nanogel in anhydrous DMSO and DMF at 50°C.

3.5. Molecular Simulation of Enzyme Nanogels as an Aid in the Designing of Enzyme Nanogels

The enzyme–acrylamide assembly driven by H bonding between the enzyme and acrylamide moieties, which is a prerequisite in the preparation of enzyme nanogels, is shown in **Fig. 10.3**. The results are confirmed by the fluorescence resonance energy transfer (FRET) spectrum (23). The root mean square distance (RMSD) value is a statistical index of the conformational similarity. The RMSDs of the backbone of the encapsulated lipase in anhydrous DMSO at 60°C in nanogels of different thicknesses are shown in **Fig. 10.4**. It can be seen that the increase in the thickness of the nanogel reduces the increase in the RMSD value, i.e., the stability of the encapsulated enzyme improves despite the denaturing effect of DMSO. Thus, molecular simulation can be used as an aid in the designing of enzyme nanogels.

As an example, MD simulation of a lipase enzyme nanogel is carried out as follows:

1. The structure of native *C. rugosa* lipase is obtained from the Brookhaven Protein Data Bank (PDB code: 1TRH) (*see* **Note 11**).
2. The GROMACS 3.3 package is used to perform MD simulation. This package is a collection of programs and libraries for MD simulation and subsequent analysis of trajectory data.

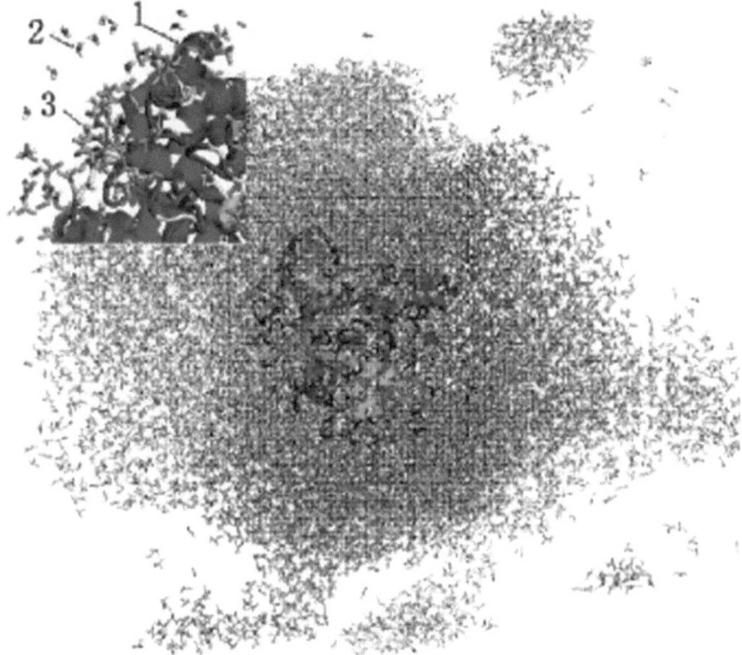

Fig. 10.3. Molecular simulation of the lipase–acrylamide assembly; *1*, lipase; *2*, water; and *3*, acrylamide [reproduced from (23) with permission from American Chemical Society (17)].

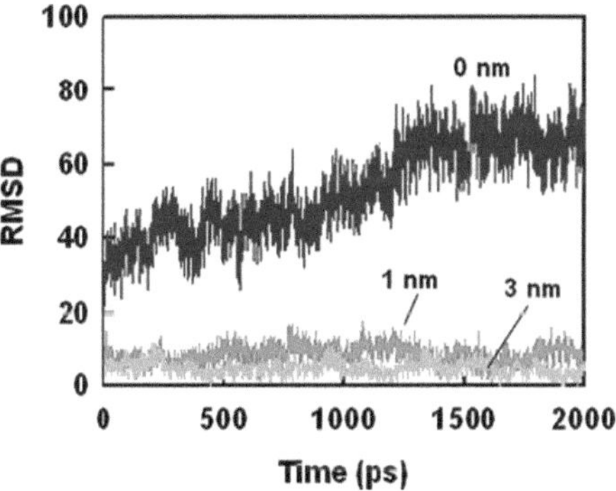

Fig. 10.4. RMSDs of native lipase and the lipase incorporated in nanogels with gel layers of thickness 1 and 3 nm at 60°C in DMSO.

3. Simulations are performed using general triclinic cell geometry. Pressure and temperature coupling is implemented for all types of simulation cells. Berendsen's weak coupling algorithm is used for both pressure and temperature.
4. The CRL molecule is placed in the center of a rectangular box using periodic boundary conditions with either water or a water/acrylamide solution. The size of the simulation box is 16.48 nm × 16.445 nm × 14.773 nm.

5. The system containing a lipase molecule and a certain number of water molecules or a water/acrylamide mixture is submitted to 500 steps of steepest descent minimization converging to a value of 2,000 kJ mol/nm by applying the particle-mesh Ewald method at 303 K (25°C).
6. A 10-ps position-restrained MD simulation is performed by keeping the protein coordinates fixed and allowing the water and acrylamide molecules to equilibrate themselves.
7. A 1000-ps MD simulation is performed at 303 K, and a leapfrog algorithm is used for integrating the Newtonian equations of motion for 500,000 simulation steps with a time step of 0.002 ps.

4. Notes

1. Deactivation of the enzyme during chemical modification often occurs due to the irreversible modification of active sites or necessary amino acid residues. Addition of a substrate or a reversible inhibitor is an effective method for protecting the active sites from unexpected modification.
2. Commercial lipase from *C. rugosa* (type VII) (CRL) purchased from Sigma-Aldrich contains insoluble components that must be removed prior to use.
3. NAS is insoluble in water and is therefore dissolved in DMSO prior to application to the aqueous phase. NAS is prone to hydrolysis and should therefore be handled under dry conditions and stored at −20°C.
4. Fluorescence detection is used to monitor the chromatographic behavior of the enzyme nanogel because polyacrylamide absorbs at 215 and 280 nm.
5. Triton X-100 is used to dissolve *p*-nitrophenyl palmitate in aqueous solution (27).
6. Parameters such as the reaction pH and NAS:enzyme molar ratio should be optimized to obtain acryloylated enzyme with high residual activity. A high NAS:enzyme molar ratio and a pH value above 9.0, which favors the unprotonated state of the primary amine group of lysine, result in a high degree of modification; however, these conditions may also promote unexpected enzyme denaturation.
7. *N,N′*-methylenebisacrylamide can be used as a cross-linker to obtain a polymer network around the enzyme. Polymerization can also be carried out without *N,N′*-methylenebisacrylamide. In this case, the acryloylated

enzyme with multiple double bonds can serve as the cross-linker. The acrylamide concentration should be optimized to obtain satisfactory encapsulation.

8. The average diameter of the HRP nanogel is 11 nm. Considering the size of the native HRP molecule, each nanogel contains only one enzyme molecule. The sizes of the HRP, BCA, and CRL molecules are 6, 5, and 7 nm (data from the Protein Data Bank), respectively.
9. Since the radius of curvature of the AFM tip is 20 nm, the diameter of the BCA nanogel is estimated to be 13 nm.
10. The diluted enzyme nanogel sample is placed on the glass surface to form a very thin film.
11. For consistency with the experimental conditions, none of the aspartates is protonated.

Acknowledgments

We gratefully acknowledge the support from the National High-tech R&D Program (863 Program; project number 2008AA05Z406) and National Natural Science Foundation (project number 20776076). The authors extend their thanks to Prof. Yunfeng Lu at Department of Chemical and Biomolecular Engineering, University of California, Los Angeles, for his helps and suggestions on the research into nanostructures.

References

1. Kim, J., Grate, J. W., and Wang, P. (2008) Nanobiocatalysis and its potential applications. *Trends Biotechnol.* **26**, 639–646.
2. Kim, J., Grate, J. W., and Wang, P. (2006) Nanostructures for enzyme stabilization. *Chem. Eng. Sci.* **61**, 1017–1026.
3. Klibanov, A. M. (2001) Improving enzymes by using them in organic solvents. *Nature* **409**, 241–246.
4. Schmid, A., Dordick, J. S., Hauer, B., Kiener, A., Wubbolts, M., and Witholt, B. (2001) Industrial biocatalysis today and tomorrow. *Nature* **409**, 258–268.
5. Xiao, Y., Patolsky, F., Katz, E., Hainfeld, J. F., and Willner, I. (2003) "Plugging into enzymes": Nanowiring of redox enzymes by a gold nanoparticle. *Science* **299**, 1877–1881.
6. Vriezema, D. M., Aragonès, M. C., Elemans, J. A. A. W., Cornelissen, J. J. L. M., Rowan, A. E., and Nolte, R. J. M. (2005) Self-assembled nanoreactors. *Chem. Rev.* **105**, 1445–1489.
7. Wu, L., and Payne, G. F. (2004) Biofabrication: Using biological materials and biocatalysts to construct nanostructured assemblies. *Trends Biotechnol.* **22**, 593–599.
8. Rosi, N. L., and Mirkin, C. A. (2005) Nanostructures in biodiagnostics. *Chem. Rev.* **105**, 1547–1562.
9. Stephanopoulos, N., Solis, E. O. P., and Stephanopoulos, G. (2005) Nanoscale process systems engineering: Toward molecular factories, synthetic cells, and adaptive devices. *AIChE J.* **51**, 1858–1869.
10. Kinbara, K., and Aida, T. (2005) Toward intelligent molecular machines: Directed motions of biological and artificial molecules and assemblies. *Chem. Rev.* **105**, 1377–1400.

11. Pohorille, A., and Deamer, D. (2002) Artificial cells: Prospects for biotechnology. *Trends Biotechnol.* **20**, 123–128.
12. Tanaka, M., and Sackmann, E. (2005) Polymer-supported membranes as models of the cell surface. *Nature* **437**, 656–663.
13. Zaks, A., and Klibanov, A. M. (1984) Enzymatic catalysis in organic media at 100°C. *Science* **224**, 1249–1251.
14. Garza-Ramos, G., Darszon, A., Tuena de Gomez-Puyou, M., and Gomez-Puyou, A. (1989) Catalysis and thermostability of mitochondrial F1-ATPase in toluene-phospholipid-low-water systems. *Biochemistry* **28**, 3177–3182.
15. Volkin, D. B., Staubli, A., Langer, R., and Klibanov, A. M. (1991) Enzyme thermoinactivation in anhydrous organic solvents. *Biotechnol. Bioeng.* **37**, 843–853.
16. Song, J. K., and Rhee, J. S. (2001) Enhancement of stability and activity of phospholipase A1 in organic solvents by directed evolution. *Biochim. Biophys. Acta* **1547**, 370–378.
17. Zhong, Z., Liu, J. L. C., Dinterman, L. M., Finkelman, M. A. J., Mueller, W. T., Rollence, M. L., Whitlow, M., and Wong, C. H. (1991) Engineering subtilisin for reaction in dimethylformamide. *J. Am. Chem. Soc.* **113**, 683–684.
18. Knubovets, T., Osterhout, J. J., and Klibanov, A. M. (1999) Structure of lysozyme dissolved in neat organic solvents as assessed by NMR and CD spectroscopies. *Biotechnol. Bioeng.* **63**, 242–248.
19. Yan, M., Ge, J., Liu, Z., and Ouyang, P. (2006) Encapsulation of single enzyme in nanogel with enhanced biocatalytic activity and stability. *J. Am. Chem. Soc.* **128**, 11008–11009.
20. Ge, J., Lu, D., Wang, J., and Liu, Z. (2009) Lipase nanogel catalyzed transesterification in anhydrous dimethyl sulfoxide. *Biomacromolecules* **10**, 1612–1618.
21. Kim, J., and Grate, J. W. (2003) Single-enzyme nanoparticles armored by a nanometer-scale organic/inorganic network. *Nano Lett.* **3**, 1219–1222.
22. Yan, M., Liu, Z., Lu, D., and Liu, Z. (2007) Fabrication of single carbonic anhydrase nanogel against denaturation and aggregation at high temperature. *Biomacromolecules* **8**, 560–565.
23. Ge, J., Lu, D. N., Wang, J., and Liu, Z. (2008) Molecular fundamentals of enzyme nanogels. *J. Phys. Chem. B* **112**, 14319–14324.
24. Davis, J. C., and Averill, B. A. (1981) Isolation from bovine spleen of a green heme protein with properties of myeloperoxidase. *J. Biol. Chem.* **256**, 5992–5996.
25. Pocker, Y., and Stone, J. T. (1965) The catalytic versatility of erythrocyte carbonic anhydrase. The enzyme-catalyzed hydrolysis of *p*-nitrophenyl acetate. *J. Am. Chem. Soc.* **87**, 5497–5498.
26. Pocker, Y., and Stone, J. T. (1967) The catalytic versatility of erythrocyte carbonic anhydrase. III. Kinetic studies of the enzyme-catalyzed hydrolysis of *p*-nitrophenyl acetate. *Biochemistry* **6**, 668–678.
27. López, N., Pernas, M. A., Pastrana, L. M., Sánchez, A., Valero, F., and Rúa, M. L. (2004) Reactivity of pure Candida rugosa lipase isoenzymes (Lip1, Lip2, and Lip3) in aqueous and organic media. Influence of the isoenzymatic profile on the lipase performance in organic media. *Biotechnol. Prog.* **20**, 65–73.

Chapter 11

Fabrication and Characterization of Bioactive Thiol-Silicate Nanoparticles

Frances Neville and Paul Millner

Abstract

Here we describe a new method for the production of thiol-silicate particles and the entrapment of enzymes within the thiol particles as they are formed. When bio-inspired polymers (polyethyleneimine) are combined with a silicic acid source and phosphate buffer under pH neutral conditions, formation of silicate particles occurs. In the method presented here the silica source contains a thiol group and so therefore the silicate particles are pre-functionalized with thiol groups. We have termed the silicate particles produced "thiol particles" and the characterization of these thiol particles is also presented in this chapter. As enzymes can be entrapped during fabrication, it means that the thiol particles can not only attach to metal surfaces but also catalyse certain reactions depending on the enzyme used. This means that there are many future possibilities for the use of thiol particles containing enzymes, as they may be used in a wide range of processes and devices which require catalytic functionalized surfaces, such as biosensors and biocatalytic reactors.

Key words: Silicates, thiol, nanoparticles, biomimetic silica, polyethyleneimine, biosensors, biocatalysis.

1. Introduction

One method of silicate nanoparticle fabrication which has been used so far is one which is based on natural biosilication (1–8). In vitro specific proteins called silaffins can catalyse silication, which occurs at near-neutral pH and ambient temperature and produces silicate nanoparticles and nanostructures (1, 2). Recent work has demonstrated that biomimetic polymers such as polyethyleneimines (PEIs) (3, 4) are also efficient at producing silicate formation of nanoparticles in a similar manner to

P. Wang (ed.), *Nanoscale Biocatalysis*, Methods in Molecular Biology 743,
DOI 10.1007/978-1-61779-132-1_11, © Springer Science+Business Media, LLC 2011

silaffin peptides (5, 6). The silicate nanoparticles and nanostructures produced using PEI can be used to entrap enzymes for a number of applications in the fields of biosensors and biocatalysis. This silication method has great potential as it may be used by many industrial sectors and it has the major advantage that it does not use harsh chemical conditions or processing and may be used in manufacturing products from pilot to bulk scale (5, 7, 8).

There are several other methodologies for the production of silicate nanoparticles. However, most are based on the Stöber process (9) in which alkoxysilanes are mixed with alcohol and polymerized in the presence of ammonia. This method has been used effectively for many years. However, it involves heating the particles to temperatures well above ambient temperature and also the use of highly basic pH (9, 10). These conditions are highly damaging to proteins and other biomolecules.

As well as the method of nanoparticle fabrication, the method in which nanoparticles are functionalized is an important issue. Most nanoparticle functionality is carried out by surface modification of nanoparticles (11). However, surface modifications have the drawback that they usually require complicated processing steps and surface coverage is low, especially when using silica nanoparticles as substrates (11). Work by Lee et al. (11) describes a method of producing silicate particles which are modified with thiol groups. However, this technique (11) requires many hours to progress and the size of the particles produced is significantly greater than that produced using the method presented in this chapter. Moreover, their methodology does not control the pH of the system which is very important when considering immobilization or entrapment of biological molecules within nanoparticles, as the biological molecules may require more neutral and ambient conditions than might otherwise be used.

A technique capable of producing silicate particles which contain biological components, such as enzymes entrapped within them, is advantageous in a number of areas (5, 7, 8). In addition, it is advantageous if the particles could also be pre-functionalized with reactive groups such as free thiol groups and give similar results, independent of the enzyme type utilized. This would enable an approach that could be more widely applied to biosensor, biocatalytic and other biofunctionalized surfaces and materials. Furthermore, if such a methodology could also be created offering greater ease of use and speed than previously used, it would represent a real innovation over previous methods for the reasons mentioned above.

In this chapter, a new one-pot method of fabrication of thiol particles and their subsequent characterization is presented. The whole fabrication process has important advantages in that it is very rapid, when compared to other recently described methods

(11) and, as well as being within the nanometre size range, the particles are pre-functionalized and require no further surface modification. Additionally, the method has the advantage that it is carried out under neutral pH conditions and at room temperature, without the use of ammonia or alcohol which are often used to produce silicate nanoparticles. The thiol particles are highly useful in producing self-assembled and biomimetic systems as they can be easily tethered to metal surfaces or metal particles. Moreover, they have the possibility of incorporating enzymatic activity within them by immobilizing enzymes within the particles or by surface immobilization via available reactive thiol groups.

Here the procedure for the production of thiol particles and thiol particles containing β-galactosidase enzyme (8) and glucose oxidase (5, 12) is presented. However, in addition, acetylcholinesterase has also been entrapped using the same method (13). All of these enzymes are commonly used in the biosensor and biocatalysis fields and they have been successfully entrapped within thiol particles using this new method (13). The characterization method which involves colorimetric assays for thiol group and enzyme activity and particle size and morphology characterization by field emission gun scanning electron microscopy and Nanosight nanoparticle tracking analysis is also presented.

2. Materials

All materials were purchased from Sigma-Aldrich, Poole, UK, unless otherwise stated. All materials should be as highly pure as possible (95–99.9% purity).

2.1. Thiol Particle Fabrication

1. *Phosphate buffer*: 290 mM (10×) Sodium phosphate buffer solution made with mono- and disodium salts, pH 7.4. Filter sterilize with 0.2 μM filters to remove contaminants. May be stored for several months at room temperature as long as no contamination is observed.
2. *Polyethyleneimine solution*: 1 mM (10×) 25-kDa polyethyleneimine (PEI) solution made with water (*see* **Note 1**). This solution may be reused and stored at room temperature for 1–2 months, although ideally it should be made fresh.
3. *3-mercaptopropyltrimethoxysilane solution*: 1 M 3-mercaptopropyltrimethoxysilane (3-mPTMOS) containing 1 mM hydrochloric acid. Mix using vortex mixing; the solution will appear cloudy but will go transparent after hydrolysis has occurred. This solution should be made fresh each time and incubated at room temperature for hydrolysis to occur.

For consistency the same hydrolysis time should be used for all experiments, generally this is 15–20 min. 3-mPTMOS is harmful and should be used in a fume hood as it has a strong smell when neat. HCl is corrosive.

2.2. Thiol Particle Fabrication with Enzyme Entrapped

1. *Phosphate buffer*: 290 mM (10×) Sodium phosphate buffer solution, pH 7.4.
2. *Polyethyleneimine solution*: 1 mM (10×) 25-kDa polyethyleneimine solution made with water.
3. *PIPES buffer*: 72.6 mM (4×) piperazine-*N*,*N*′-bis(2-ethanesulphonic acid) (PIPES) buffer solution, pH 7.4. Filter sterilize with 0.2 μM filters to remove contaminants. May be stored for several months at room temperature as long as no contamination is observed.
4. *β-galactosidase enzyme solution*: 10 mg/mL β-Galactosidase (type VI from *Escherichia coli*) diluted in water. Make fresh and store on ice or at 4°C.
5. *Glucose oxidase enzyme solution*: 10 mg/mL glucose oxidase diluted in water. Make fresh and store on ice or at 4°C.
6. *3-mercaptopropyltrimethoxysilane solution*: 1 M 3-mercaptopropyltrimethoxysilane (3-mPTMOS) containing 1 mM hydrochloric acid.

2.3. Thiol Particle Characterization – Thiol Groups and Enzyme Activity

2.3.1. Ellman's Reagent Test

1. *Phosphate buffer*: 0.1 M sodium phosphate buffer, pH 7.0.
2. *Ellman's reagent solution*: 10 mM Ellman's reagent (5,5′-dithio-bis(2-nitrobenzoic acid)) made up in 0.1 M sodium phosphate, pH 7.0 (14). This solution should be made fresh each time and kept on ice.
3. *Thiol particle solution*: Thiol particles with or without enzyme, diluted in water.
4. *3-mercaptopropyltrimethoxysilane solution*: Solution of 0.1 M 3-mPTMOS hydrolysed with 1 mM HCl.

2.3.2. ONPG Test (β-Galactosidase Activity)

1. *Phosphate buffer*: 0.1 M sodium phosphate.
2. *ONPG solution*: A solution of approximately 20.5 mg/mL *ortho*-nitrophenyl-β-D-galactopyranoside (ONPG) made up in 0.1 M sodium phosphate buffer, pH 7.4. The solution should be gently warmed to completely dissolve the product. This solution should be made fresh each time.
3. *Thiol particle solution*: Solution of thiol particles prepared using β-galactosidase diluted in water.
4. *β-galactosidase enzyme solution*: 1 mg/mL β-galactosidase (type VI from *E. coli*) diluted in water. Make fresh and store on ice or at 4°C.

2.3.3. Glucose Oxidase Activity

1. *HRP solution*: 25 units/mL horseradish peroxidase. This solution should be made fresh each time and stored on ice.
2. *ABTS solution*: 25 mM 2,2′-azino-bis(3-ethylbenzthiazoline-6-sulphonic acid) (ABTS). This solution should be made fresh each time and stored on ice or at 4°C
3. *Phosphate buffer*: 0.1 M sodium phosphate, pH 7.0.
4. *Glucose solution*: 1 M glucose solution. This solution should be made at least 1 day in advance so that the D- and L-forms can equilibrate. The solution should be filter sterilized as it can easily become contaminated.
5. *Thiol particle solution*: Solution of thiol particles prepared using glucose oxidase diluted in water.
6. *Glucose oxidase enzyme solution*: 1 mg/mL glucose oxidase diluted in water. Make fresh and store on ice or at 4°C.

2.4. Thiol Particle Characterization – Physical and Chemical Properties

In order to determine the particle size distribution, the particles in suspension may be sized by analysis of Brownian motion using video microscopy (15): the Nanosight NTA system employs laser light scattering to enable particle motion tracking (16). After video capture and analysis is carried out, the hydrodynamic diameter of the particles is calculated by the Nanosight software using the Stokes–Einstein equation (16):

2.4.1. Particle Size Distribution – Nanosight Nanoparticle Tracking Analysis System

1. Thiol particles with or without enzyme, diluted in water. A range of dilutions should be prepared. The amount of particles which form may vary slightly from batch to batch. Therefore, make several dilutions of particles using a serial dilution. As a general guide, for use with a Nanosight nanoparticle tracking analysis (NTA) system, if at all the solution looks cloudy, it is far too concentrated. The concentration of particles can affect the particle analysis. If there are too many particles, a clear video cannot be taken, whilst too few particles may not be representative of the whole sample.
2. Ethanol (for cleaning).
3. Water (for cleaning).

2.4.2. Field Emission Gun Scanning Electron Microscopy (FEGSEM) – Particle Solution Characterization

1. Thiol particles with or without enzyme, diluted in water. In this case the particle suspension should be more concentrated than that used for Nanosight NTA. If the solution is concentrated, the particles will form surface-covering self-assembled structures which will reduce in size if the particles are more diluted, leaving some areas of the surfaces uncovered. The concentration used will depend on the desired application. However, in general a range of concentrations should be used to characterize the particles.
2. Aluminium mounting stubs (Agar Scientific, Stansted, UK).

2.4.3. FEGSEM – Thiol Particle Binding to Gold Surfaces

1. Thiol particles with or without enzyme, diluted in water.
2. Gold electrodes. In the authors' case, these were obtained from the Tyndall Institute, Cork, and comprised 200 nm gold onto a 50-nm titanium adhesion layer on SiO_2.
3. Hydrogen peroxide (corrosive).
4. Sulphuric acid (corrosive).
5. Aluminium mounting stubs (Agar Scientific, UK).
6. Conducting carbon tape (Agar Scientific, UK).
7. Conducting carbon paint (Agar Scientific, UK).

2.4.4. FEGSEM – Blocking Thiol Particle-Specific Binding to Gold Surfaces via Alkylation

1. Thiol particles with or without enzyme, diluted in water.
2. 0.1 M sodium phosphate buffer at pH 6.5.
3. 20 mM *N*-ethylmaleimide (NEM) in 0.1 M phosphate buffer at pH 6.5. The use of this pH prolongs the half-life of the NEM which means that the solution and reaction will be stable for a longer time period than the length of time the experiment takes to perform. This solution should be made fresh for each experiment.
4. Gold electrodes. In the authors' case these were obtained from the Tyndall Institute, Cork, and comprised 200 nm gold onto a 50-nm titanium adhesion layer on SiO_2.
5. Hydrogen peroxide (corrosive).
6. Sulphuric acid (corrosive).
7. Aluminium mounting stubs (Agar Scientific, UK).
8. Conducting carbon tape (Agar Scientific, UK).
9. Conducting carbon paint (Agar Scientific, UK).

3. Methods

3.1. Thiol Particle Fabrication

1. Prepare all the solutions stated in **Section 2.1**, preparing the 3-mPTMOS last. Carry out the following steps during the 15–20 min hydrolysis incubation so that the hydrolysed 3-mPTMOS can be added at the correct time point.
2. Add 0.7 mL water to a 1.5-mL microcentrifuge tube.
3. To this add 0.1 mL 290 mM (10X) sodium phosphate buffer, pH 7.4, and mix.
4. To this add 0.1 mL 1 mM PEI (10X) and mix.
5. After 15–20 min hydrolysis time (the 3-mPTMOS solution should go transparent), add 0.1 mL hydrolysed 1 M

3-mPTMOS to the microcentrifuge tube (total volume 1 mL) and vortex mix.

6. Leave the solution to incubate at room temperature.
7. After a period of time (5–15 min, depending on the ambient room temperature) the solution will go opaque due to the formation of thiol particles.
8. At this point, wash the particles by centrifuging at 16,873×*g* for 5 min, removing the supernatant and then resuspending in water. Repeat this process so that the particles have been washed twice. After the second wash, resuspend the particles in 0.5 mL water for storage and characterization (*see* **Note 2**).

3.2. Thiol Particle Fabrication with Enzyme Entrapped

1. Prepare all the solutions stated in **Section 2.2**, preparing the 3-mPTMOS last.
2. Label two microcentrifuge tubes "A" and "B".
3. To tube "A" add 0.2 mL water, 0.1 mL 290 mM sodium phosphate buffer, pH 7.4, and 0.1 mL 1 mM PEI. Leave to incubate for 15 min.
4. Concurrently, add 0.4 mL 72.6 mM PIPES buffer and 0.1 mL enzyme (10 mg/mL β-galactosidase or glucose oxidase) to tube "B" and mix (*see* **Note 3**).
5. After 15–20 min hydrolysis time (the 3-mPTMOS solution should go transparent), add 0.1 mL hydrolysed 1 M 3-mPTMOS to tube "B" (total volume 0.6 mL) and vortex mix. Leave tube "B" to incubate at room temperature for 15 min.
6. After both tubes have incubated for 15 min, add the contents of both tubes together in one tube. Then vortex mix and incubate at room temperature.
7. After a period of time (5–15 min, depending on the ambient room temperature), the solution will go opaque due to the formation of thiol particles with enzyme. For a flowchart of this method, *see* **Flowchart 11.1**.
8. At this point the particles should be washed by centrifuging at 16,873×*g* for 2 min, removing the supernatant and then resuspending in water (*see* **Note 4**). Repeat this process so that the particles have been washed twice. After the second wash, resuspend the particles in 0.5 mL water for storage and characterization (*see* **Note 5**).

3.3. Thiol Particle Characterization – Thiol Groups and Enzyme Activity

After synthesis of the thiol particles with or without enzyme entrapped, the particles may be tested in different ways in order to ensure that the thiol particles contained free thiol groups and were enzyme active.

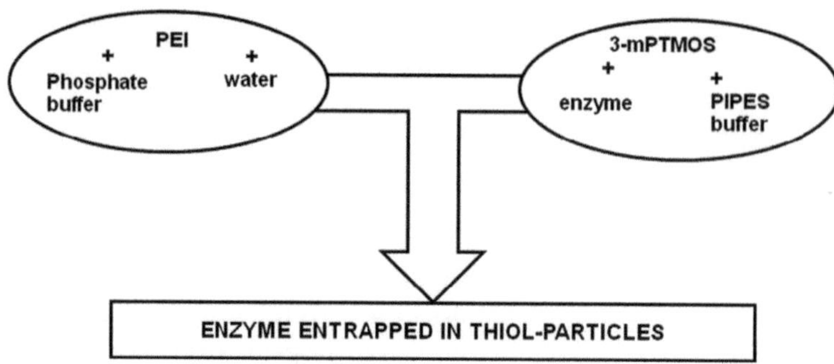

Flowchart 11.1. In order to entrap enzymes within the thiol particles, PEI and phosphate are incubated together in one tube, whilst hydrolysed 3-mPTMOS, PIPES buffer and the enzyme are incubated in a separate tube. After 15 min, the contents of both tubes are added together and incubated for a further 5–15 min until thiol particles containing enzyme form.

3.3.1. Ellman's Test

The first test uses Ellman's reagent (14) (5,5′-dithio-bis(2-nitrobenzoic acid)) which reacts with free thiol groups to produce yellow thio-nitrophenol. The presence of a yellow colour shows that there are reactive free thiol groups present on the surface of the particles which could be used for attachment of the particles to metal surfaces:

1. Label three microcentrifuge tubes: "positive", "negative" and "sample".
2. Add 0.1 mL 10 mM Ellman's reagent (5,5- dithio-bis(2-nitrobenzoic acid)) made up in 0.1 M sodium phosphate, pH 7.0, to all three tubes.
3. In addition, add 0.9 mL water to the "negative" tube and mix.
4. To tube "positive" also add 0.8 mL water and 0.1 mL 0.1 M hydrolysed 3-mPTMOS and mix.
5. To the "sample" tube, add 0.7 mL water and 0.2 mL thiol particles diluted in water and mix (*see* **Note 6**).
6. Leave the three tubes to incubate for 30 min.
7. For qualitative results, compare the yellow colour produced by the three tubes. The "positive" tube should give a strong dark yellow colour, the "negative" a very pale yellow colour and the "sample" tube should be a darker yellow colour than the "negative". The colour produced by the sample will depend on the dilution of the thiol particles.
8. For quantitative results, produce a standard curve using known concentrations of thiol groups and read the yellow colour at a certain time point at 405 nm on a spectrophotometer. Then test the absorption of the sample at 405 nm at the same time point. Note that the sample should be centrifuged to remove the particles prior to testing so that

there is no interference from the particulate matter when the absorbance is read. Also test a negative control and subtract this as the background reading.

3.3.2. ONPG Test (β-galactosidase Activity)

β-galactosidase activity can be measured using *ortho*-nitrophenyl-β-D-galactopyranoside (ONPG) which is converted to yellow *ortho*-nitrophenol and galactose by the β-galactosidase enzyme:

1. Label three microcentrifuge tubes: "positive", "negative" and "sample".
2. Add 0.1 mL ONPG made up in 0.1 M sodium phosphate, pH 7.4, to all three tubes.
3. In addition, add 0.9 mL water to the "negative" tube and mix.
4. To tube "positive", also add 0.8 mL water and 0.1 mL 1 mg/mL β-galactosidase and mix.
5. To the "sample" tube, add 0.7 mL water and 0.2 mL thiol particles made with β-galactosidase enzyme diluted in water and mix (*see* **Note 6**).
6. Leave the three tubes to incubate for 30 min.
7. For qualitative results, compare the yellow colour produced by the three tubes. The "positive" tube should give a strong dark yellow colour, the "negative" tube should give a very pale yellow colour and the "sample" tube should be a darker yellow colour than the "negative". The colour produced by the sample will depend on the dilution of the thiol particles and may continue to develop with time.
8. For quantitative results, produce a standard curve using known concentrations of β-galactosidase enzyme and read the yellow colour at a certain time point at 405 nm on a spectrophotometer. Then test the absorption of the sample at 405 nm at the same time point. Note that the sample should be centrifuged to remove the particles prior to testing so that there is no interference from the particulate matter when the absorbance is read. Also test a negative control and subtract this as the background reading.

3.3.3. Glucose Oxidase Activity

The glucose oxidase assay is based on the following reactions, where glucose oxidase converts glucose, oxygen and water to gluconolactone and hydrogen peroxide. The hydrogen peroxide then oxidizes the 2,2′-azino-bis(3-ethylbenzthiazoline-6-sulphonic acid) (ABTS) and produces a blue/green colour:

$$\beta\text{-}D\text{-Glucose} + O_2 + H_2O \xrightarrow{\text{glucose oxidase}} D\text{-Glucono-1,5-Lactone} + H_2O_2$$

$$H_2O_2 + \text{ABTS (reduced - colourless)} \xrightarrow{\text{horseradish peroxidase}} \text{ABTS (oxidised – green/blue)}$$

1. Label three microcentrifuge tubes: "positive", "negative" and "sample".
2. Prepare a solution of 2.8 mL 0.1 M sodium phosphate, 0.4 mL 25 units/mL horse radish peroxidase and 0.4 mL 25 mM ABTS. This solution will look a grey/brown colour.
3. Add 0.8 mL of the solution made in Step 2 to all three tubes.
4. In addition, add 0.1 mL water to the "negative" tube and mix.
5. To tube "positive", add 0.1 mL 1 mg/mL glucose oxidase and mix.
6. To the "sample tube", add 0.1 mL thiol particles made with glucose oxidase enzyme diluted in water and mix (*see* **Note 6**).
7. Add 0.1 mL 1 M glucose to all three tubes and mix.
8. For qualitative results, compare the green/blue colour produced by the three tubes. The "positive" tube should give a strong dark turquoise colour, the "negative" tube should give a very pale turquoise colour and the "sample" tube should be a darker turquoise colour than the "negative". The colour produced by the sample will depend on the dilution of the thiol particles and may continue to develop with time (*see* **Note 7**).

3.4. Thiol Particle Characterization – Physical and Chemical Properties

3.4.1. Particle Size Distribution – Nanosight Nanoparticle Tracking Analysis System

The Nanosight NTA system is used in order to determine the hydrodynamic diameter of the particles (15, 16):

1. Set up the Nanosight NTA system.
2. Clean the cell thoroughly with ethanol.
3. Rinse the cell with water and observe the water sample to ensure the cell is clean before commencing sample analysis.
4. Test each sample by injecting approximately 0.5 mL sample into the cell, capturing the video and analysing it. Be careful not to introduce air into the sample cell.
5. Make sure to repeat Steps 2 and 3 before analysing a new sample.
6. Analyse different concentrations of the same sample to obtain the most appropriate video for subsequent analysis (*see* **Note 8**).

3.4.2. FEGSEM – Particle Solution Characterization

1. Obtain the appropriate aluminium mounting stubs for the FEGSEM.
2. Place the stubs in a stub holder or other appropriate container so that they do not move and the surface is horizontal.
3. Label the stubs so that you can identify which sample is on which stub.

4. Place a small drop of the particle suspension onto the aluminium stub and leave to dry. You can either place a large drop (0.2 mL) on one stub or divide the surface of the stub into sections and use small drops (0.01 mL) to have several samples on stub. This may increase the number of samples that can be analysed if time on the FEGSEM is limited.
5. Once the samples have thoroughly dried, use an air spray to remove any particles which are not attached to the surface. If this is not carried out, the sample may enter the FEGSEM and cause problems to occur.
6. Then place the samples in a sputter coater and sputter with 3–5 nm thickness platinum/palladium.
7. Analyse the samples using the FEGSEM. Note down the magnification and any observations made. Save images of each sample which may be later analysed for particle morphology and size distribution (*see* **Note 8**). Typical images of thiol particles with and without enzyme are seen in **Fig. 11.1**.

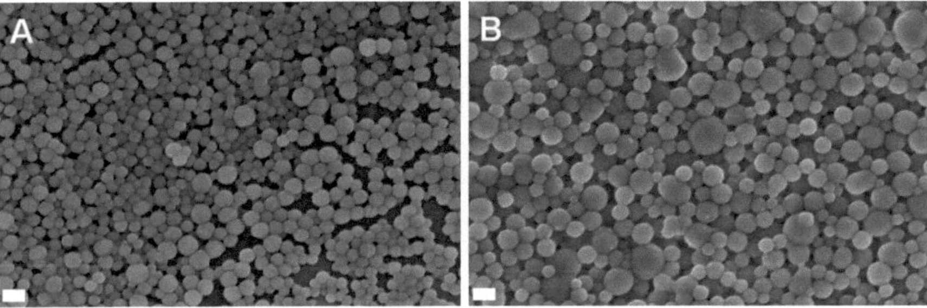

Fig. 11.1. Thiol particles were imaged using a LEO1530 Gemini FEGSEM. The particles were dried directly onto aluminium stubs. (**a**) Freshly prepared thiol particles from a water suspension were imaged using FEGSEM and the relative concentration of the suspension was ~1 mg/mL. (**b**) Thiol particles containing β-galactosidase were also imaged using FEGSEM at the same relative concentration. When the particles contain enzyme, there is a bimodal distribution of particle sizes, where the larger particles contain the majority of the enzyme introduced into the entrapment reaction. The scale bar is 200 nm.

3.4.3. FEGSEM – Thiol Particle Binding to Gold Surfaces

The thiol particles were incubated with gold electrodes to see if they would bind specifically to gold surfaces via the thiol groups on their surfaces:

1. Prepare piranha solution using 30% (v/v) hydrogen peroxide and 70% (v/v) sulphuric acid. This should be carried out in a fume hood using protective clothing, safety glasses and gloves. The solution should be made in a beaker within a secondary container such as a larger beaker with a glass plate on top. Take care as the solution is extremely corrosive.
2. Thoroughly clean gold electrodes using fresh piranha solution by immersing the electrodes in the piranha solution for 2 min. Use nylon forceps so that the electrodes do

not become scratched and are not damaged by the piranha solution.

3. Then gently rinse the electrodes with water before drying them using a stream of nitrogen.
4. Incubate solutions of thiol particles with the clean gold electrodes for 1 h.
5. Then rinse the electrodes with water and dry with a stream of nitrogen.
6. Mount the electrodes onto FEGSEM aluminium stubs for imaging by using conducting tape.
7. Paint a thin layer of carbon-conducting paint around the edge of the electrode so that the electrode is in contact with the aluminium stub.
8. Analyse the samples using the FEGSEM. Note down the magnification and any observations made. Save images of each sample which may be later analysed for particle morphology and size distribution (*see* **Note 8**).

3.4.4. FEGSEM – Blocking Thiol Particle-Specific Binding to Gold Surfaces

As well as directly observing the particles binding to the gold surface, it is possible to confirm that the binding is specific by blocking the free thiol groups using *N*-ethylmaleimide (NEM) which alkylates thiol groups. The use of NEM for blocking free thiol groups for enzyme inhibition assays has long been used (17, 18), with NEM covalently binding to thiol groups at pH 6.5–7.5 (19):

1. Add equal parts of 20 mM *N*-ethylmaleimide (NEM) in 0.1 M phosphate buffer at pH 6.5 and thiol particle solutions to give a final NEM concentration of 10 mM and mix.
2. Incubate the particles and NEM for 1 h at room temperature.
3. During the incubation period, carry out electrode preparation steps as in Steps 1–3 of **Section 3.4.3**.
4. If desired you may test the samples before and after for thiol group presence (to confirm NEM is blocking thiol groups). Use the Ellman's test for this (**Section 3.3.1**) but have two sample tubes (before and after NEM treatment).
5. Incubate the NEM-treated thiol particles for 1 h with the cleaned gold electrodes.
6. Then carry out mounting and analysis as in Steps 5–8 of **Section 3.4.3**.
7. Compare the same sample before and after NEM treatment. There should not be any binding of the thiol particles to the gold electrode surface after NEM treatment. *See* **Fig. 11.2** for an example of typical results.

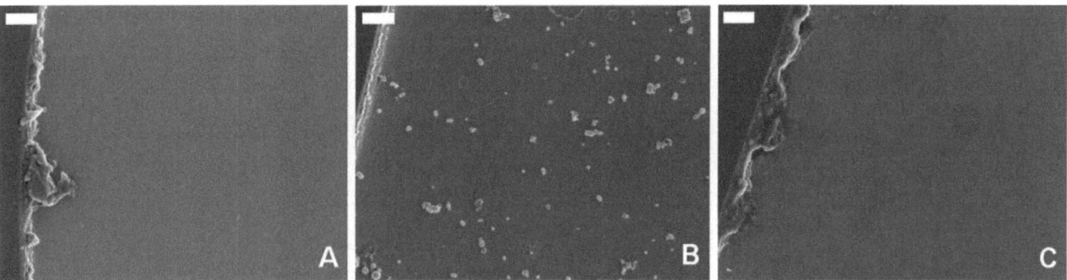

Fig. 11.2. LEO1530 Gemini FEGSEM images of (**a**) clean gold surface; (**b**) gold electrode after incubation with 3-mPTMOS particles containing β-galactosidase, showing 3-mPTMOS particles binding to gold; and (**c**) gold electrode after incubation with NEM-treated thiol particles containing β-galactosidase. The difference between images (**b**) and (**c**) clearly shows that treatment with *N*-ethylmaleimide blocks the thiol groups of the thiol particles, thus preventing specific binding. The scale bar is 1 μm.

4. Notes

1. Unless stated otherwise, all solutions should be prepared in water that has a resistivity of 18.2 MΩ cm and total organic content of less than five parts per billion. This standard is referred to as "water" in this text.
2. The final concentrations of the components will be 0.1 M 3-mPTMOS, 0.1 mM PEI and 29 mM phosphate buffer to give a phosphate concentration to PEI repeat unit ratio of 0.5.
3. In this chapter, instructions have been given on how to fabricate and characterize particles containing β-galactosidase or glucose oxidase, but other enzymes may be entrapped depending on the enzymatic reaction that is required. The concentration of the enzyme may also be varied depending on the application and sensitivity that is desired.
4. The particles produced using enzyme should be of two different size distributions with those that contain enzyme being denser and larger than the smaller and less dense thiol particles which contain little, if any, enzyme (13). Therefore, when washing the particles, the more dense particles pellet at the bottom of the tube more quickly than do the smaller particles. If you do not wish to keep the smaller particles, the cloudy supernatant containing the smaller particles may be discarded during the washing process. If you wish to keep the smaller particles, it is best to remove the supernatant to another tube and centrifuge for 5 min. Then wash both sets of particles. Remember to label the tubes so that there is no confusion during washing steps.

5. The final concentrations of the components will be 0.1 M 3-mPTMOS, 0.1 mM PEI, 29 mM PIPES buffer, 29 mM phosphate buffer and 1 mg/mL enzyme.
6. Depending on the amount of particles fabricated, the volumes can be decreased in the same ratios and so the same sample can be further characterized.
7. In general thiol particles should be approximately 50–200 nm in diameter. If enzyme has been entrapped, a bimodal distribution of particle size will occur with smaller particles being 50–200 nm and larger particles being 300–500 nm. It has been proposed (13) that the larger particles contain the vast majority of the entrapped enzyme.
8. More examples of results using the methods in this chapter have been recently published (13).

Acknowledgement

The authors would like to acknowledge the sponsorship of the SANTS project by the European Commission (Project No. NMP4-CT-2006-033254).

References

1. Niemeyer, C. M. (2001) Nanoparticles, proteins, and nucleic acids: Biotechnology meets materials science. *Angew. Chem. Int. Ed.* **40**, 4128–4158.
2. Lopez, P. J., Gautier, C., Livage, J., and Coradin, T. (2005) Mimicking biogenic silica nanostructures formation. *Curr. Nanosci.* **1**, 73–83.
3. Jin, R. H., and Yuan, J. J. (2005) Simple synthesis of hierarchically structured silicas by poly(ethyleneimine) aggregates pre-organized by media modulation. *Macromol. Chem. Phys.* **206**, 2160–2170.
4. Berne, C., Betancor, L., Luckarift, H. R., and Spain, J. C. (2006) Application of a microfluidic reactor for screening cancer prodrug activation using silica-immobilized nitrobenzene nitroreductase. *Biomacromolecules* **7**, 2631–2636.
5. He, P., Greenway, G., and Haswell, S. J. (2008) The on-line synthesis of enzyme functionalized silica nanoparticles in a microfluidic reactor using polyethylenimine polymer and R5 peptide. *Nanotechnology* **19**, 315603.
6. Helmecke, O., Hirsch, A., Behrens, P., and Menzel, H. (2008) Influence of polymeric additives on biomimetic silica deposition on patterned microstructures. *J. Colloid Interface Sci.* **321**, 44–51.
7. Betancor, L., and Luckarift, H. R. (2008) Bioinspired enzyme encapsulation for biocatalysis. *Trends Biotechnol.* **26**, 566–572.
8. Betancor, L., Luckarift, H. R., Seo, J. H., Brand, O., and Spain, J. C. (2008) Three-dimensional immobilization of β-galactosidase on a silicon surface. *Biotechnol. Bioeng.* **99**, 261–267.
9. Stöber, W., Fink, A., and Bohn, E. (1968) Controlled growth of monodisperse silica spheres in the micron size range. *J. Colloid Interface Sci.* **26**, 62–69.
10. Hernandez, G., and Rodriquez, R. (1999) Adsorption properties of silica sols modified with thiol groups. *J. Non-Cryst. Solids* **246**, 209–215.
11. Lee, Y. G., Park, J. H., Oh, C., Oh, S. G., and Kim, Y. C. (2007) Preparation of highly monodispersed hybrid silica spheres using a

one-step sol-gel reaction in aqueous solution. *Langmuir* **23**, 10875–10878.

12. Pchelintsev, N. A., and Millner, P. A. (2008) A novel procedure for rapid surface functionalisation and mediator loading of screen-printed carbon electrodes. *Anal. Chim. Acta* **612**, 190–197.
13. Neville, F., Pchelintsev, N. A., Broderick, M. J. F., Gibson, T., and Millner, P. A. (2009) Novel one-pot synthesis and characterization of bioactive thiol-silicate nanoparticles for biocatalytic and biosensor applications. *Nanotechnology* **20**, 055612.
14. Ellman, G. L. (1958) A calorimetric method for determining low concentrations of mercaptans. *Arch. Biochem. Biophys.* **74**, 443–450.
15. Nakroshis, P., Amoroso, M., Legere, J., and Smith, C. (2003) Measuring Boltzmann's constant using video microscopy of Brownian motion. *Am. J. Phys.* **71**, 568–573.
16. Carr, B., and Malloy, A. (2006) Nanoparticle tracking analysis – the NANOSIGHT system. http://www.nanosightuk.co.uk/pdfs/NanoSight%20Technology.pdf
17. Roberts, E., and Rouser, G. (1958) Spectrophotometric assay for reaction of *N*-ethylmaleimide with sulfhydryl groups. *Anal. Chem.* **30**, 1291–1292.
18. Alexander, N. M. (1958) Spectrophotometric assay for sulfhydryl groups using *N*-ethylmaleimide. *Anal. Chem.* **30**, 1292–1294.
19. Nishiyama, J., and Kuninori, T. (1992) Assay of thiols and disulfides based on the reversibility of *N*-ethylmaleimide alkylation of thiols combined with electrolysis. *Anal. Biochem.* **200**, 230.

Chapter 12

Immobilization of Enzymes on Fumed Silica Nanoparticles for Applications in Nonaqueous Media

Juan C. Cruz, Kerstin Würges, Martin Kramer, Peter H. Pfromm, Mary E. Rezac, and Peter Czermak

Abstract

Enzymatic catalysis in nonaqueous media is considered as an attractive tool for the preparation of a variety of organic compounds of commercial interest. This approach is advantageous for numerous reasons including the enhanced stability of some substrates and products in solvents, sometimes improved selectivity of the enzyme, and reduction of unwanted water-dependent side reactions since little water is present. Due to the poor solubility of enzymes in these media, mass transfer limitations are sometimes present, leading to low apparent catalytic activity. Immobilization on solid supports has been successfully applied to overcome enzyme solubility issues by increasing the accessibility of substrates to the enzymes' active sites. We have developed a simple immobilization protocol that uses fumed silica as support. Fumed silica is an inexpensive nanostructured material with unique properties including large surface area and exceptional adsorptive affinity for organic macromolecules. Our protocol is performed in two main steps. First, the enzyme molecules are physically adsorbed on the surface of the non-porous fumed silica nanoparticles with the participation of silanol groups (Si–OH) and second, water is removed by lyophilization. The protocol has been successfully applied to both *s. Carlsberg* and *Candida antarctica* lipase B (CALB). The resulting fumed silica-based nanobiocatalysts of these two enzymes were tested for catalytic activity in hexane. The transesterification of *N*-acetyl-L-phenylalanine ethyl ester was the model reaction for *s. Carlsberg* nanobiocatalysts. The simple esterification of geraniol and the enantioselective transesterification of (*RS*)-1-phenylethanol were the model reactions for CALB nanobiocatalysts. The observed catalytic activities were remarkably high and even exceeded those of commercially available preparations.

Key words: Fumed silica, enzyme immobilization, CALB, subtilisin Carlsberg, nonaqueous media, hexane.

1. Introduction

The exquisite regio- and stereoselectivity of enzymes have contributed to their current popularity as catalysts for the synthesis of a variety of compounds of commercial interest including polymers

P. Wang (ed.), *Nanoscale Biocatalysis*, Methods in Molecular Biology 743,
DOI 10.1007/978-1-61779-132-1_12,

(1, 2), fragrances (3, 4), and pharmaceuticals (5). Some of these synthetic procedures are carried out in nonaqueous media such as hexane mainly due to the high solubility and stability of substrates and products, and sometimes the absence of side reactions such as hydrolysis and decomposition (6, 7). A major limitation arises from the fact that enzymes exhibit a poor catalytic performance in such media relative to aqueous systems (8). Among the many proposed strategies to overcome this limitation, immobilization on solid porous and non-porous supports has been quite successful (9, 10). The large-scale application is, however, limited mainly because of the high price of the resulting preparations (11, 12). The advent of nanotechnology has brought to light a number of nanostructured materials with unique characteristics that are now being considered for addressing this major bottleneck. We have developed a novel immobilization method that incorporates fumed silica nanoparticles as carriers for the enzyme molecules (13–15). Fumed silica is an amorphous non-porous material with a large specific surface area of up to 500 m^2/g (16). Due to the presence of negatively charged silanol groups at its surface, fumed silica provides facilitated adsorption pathways for a variety of macromolecules including polymers and proteins (17–19). These exceptional adsorptive properties are exploited in our methodology to generate highly active enzyme-based nanobiocatalysts.

The aim of this chapter is to present an overview of the studies performed in our laboratories on the development of a methodology to prepare fumed silica-based nanobiocatalysts for nonaqueous media. *Candida antarctica* lipase B (CALB) and *s. Carlsberg* are used here as they have shown practical applications in organic solvents. The immobilization strategy consists of two major sequential steps: the enzyme molecules are physically adsorbed on the nanoparticles in aqueous suspensions of fumed silica and the resulting preparation is lyophilized to produce the supported nanobiocatalysts for use in nonaqueous media. Our findings suggest that the catalytic efficiencies are essentially dependent on the surface coverage by the enzyme molecules. We discovered that *s. Carlsberg* constantly increased its catalytic activity as more surface area was provided for immobilization (13). In the case of CALB, at first the catalytic activity followed this trend but then decreased dramatically when the low surface coverage regime is reached (15). This optimum catalytic activity at an intermediate surface coverage was previously identified by Bosley and coworkers (20) for both *Rhizomucor miehei* and *Humicola* sp. lipases and more recently suggested for CALB by Gross and coworkers (11, 12). Our results seem to support the notion that enzymes with a tendency for deformation and multi-point attachment on the support surface show the best activity when some enzyme–enzyme interactions stabilize the active configuration of the enzyme vs. a deformed inactive configuration when excess surface is available. Enzymes with a weak tendency for deformation when attached to the support show optimum catalytic activity near and below the nominal monolayer surface coverage as based on X-ray crystallographic enzyme dimensions.

2. Materials

2.1. Enzyme Immobilization

1. Crude CALB (E.C. 3.1.1.3, lyophilized, specific activity of 30 U/mg solid; Biocatalytics, Inc., Pasadena, CA) stored at 4°C (*see* **Note 1**).
2. Crude *s. Carlsberg* (E.C. 3.4.21.14, proteinase from *Bacillus licheniformis*, specific activity of 8 U/mg solid; Sigma, St. Louis, MO) stored at –20°C (*see* **Note 1**).
3. Fumed silica (purity of 99.8%, specific surface area 255 m^2/g, primary particle diameter ≅ 7–50 nm; as reported by the manufacturer; Sigma, St. Louis, MO).
4. 1 M aqueous potassium hydroxide (KOH) solution in Milli-Q® water.
5. 10 mM aqueous monobasic potassium phosphate (KH_2PO_4) buffer solution in Milli-Q® water adjusted to pH 7.8 using the 1 M KOH solution (*see* **Note 2**).
6. Vortex mixer (Fisher Scientific, Pittsburgh, PA).
7. Sonicator (8892 Model; Cole Parmer, Vernon Hills, IL).
8. Lyophilizer (VirTis model 10-MR-TR; Gardiner, NY).

2.2. Initial Reaction Rates and Kinetic Parameters

The following items are common for all reactions:

1. 20–200 μL and 100–1,000 μL Finnpipettes (Fisher Scientific, Pittsburgh, PA).
2. Temperature-controlled shaker incubator (PsyCro Term; New Brunswick Scientific, Edison, NJ).

Specific materials for each reaction system are listed below:

Transesterification reactions with *s. Carlsberg*:

1. Molecular sieves (4–8 mesh beads) (Fisher Scientific, Pittsburgh, PA).
2. 15-mL glass vials (Teflon screw-capped, flat-bottom; Fisher Scientific, Pittsburgh, PA).

Esterification of geraniol with CALB:

1. 2- and 24-mL glass vials (Teflon screw-capped, flat-bottom; Fisher Scientific, Pittsburgh, PA).

Enantioselective transesterification of (*RS*)-1-phenylethanol with CALB:

1. 12-mL glass vials (open top caps and septum; National Scientific, Rockwood, TN).
2. PTFE syringe filters (0.2 μm pore size; Whatman, Inc.).
3. 2-mL glass vials (Teflon screw-capped, flat-bottom; Fisher Scientific, Pittsburgh, PA).

2.3. Chromatographic Analysis

GC analysis for the transesterification of *N*-acetyl-l-phenylalanine ethyl ester (APEE) with *s. Carlsberg*:

1. 1.5-mL Eppendorf tubes (Eppendorf, Westbury, NY).
2. Microcentrifuge (Centrific Model 228; Fisher Scientific, Pittsburgh, PA).
3. GC (Model 3800; Varian, Inc., Palo Alto, CA) equipped with a DB-5 capillary column (30 m length, 0.25 mm i.d., 0.25 μm film thickness; J&W Scientific, Folsom, CA) and a FID detector.

GC analysis for the esterification of geraniol with CALB:

1. GC (Model 3800; Varian, Inc., Palo Alto, CA) equipped with a DB-WAX capillary column (30 m length, 0.25 mm i.d.; J&W Scientific, Folsom, CA) and a FID detector.

HPLC analysis for the enantioselective transesterification of (*RS*)-1-phenylethanol with CALB:

1. HPLC system (LC-10A series; Shimadzu, Kyoto, Japan) equipped with a chiral column (chiracel OD-H, 25 cm length, 0.46 cm i.d.; Daicel Chemical Industries, Tokyo, Japan), pumps LCD-10ATvp liquid chromatography, a degasser DGU-14A, an auto injector SIL-10ADvp, a system controller SCL-10Avp, a column oven CTO-10Avp, a diode array detector SPD-M10Avp, and the Shimadzu Chromatography Laboratory Automated Software System Version 7.

3. Methods

3.1. Enzyme Immobilization

1. Weigh the as-received crude enzyme into glass vials (*see* **Note 3, Tables 12.1** and **12.2**).
2. Dissolve the enzyme by adding the KH_2PO_4 buffer. Obtain homogeneous solutions with concentrations of 0.1–6.7 mg of enzyme/mL by vortexing for 30 s (*see* **Tables 12.1** and **12.2**)
3. Add fumed silica to form preparations at various %nominal surface coverages (%SC) (*see* **Note 4**). Vortex for 2–3 min until a visually homogeneous suspension is formed. Sonicate for 10 min in a water bath at room temperature.
4. Place the preparations in a refrigerator at about –20°C until frozen.

Table 12.1
Summary of the amount of fumed silica and crude enzyme employed to prepare *s. Carlsberg*-supported nanobiocatalysts with a variety of nominal surface coverages. In all cases, the volume of buffer was 3 mL

Target preparations (%nominal surface coverage)	Crude *s. Carlsberg*[a] (mg)	Fumed silica[b] (mg)	Enzyme concentration (mg/mL)
1	0.3	29.7	0.1
2	0.3	19.7	0.1
7	1.0	19.0	0.3
10	2.0	18.0	0.7
135	10.0	10.0	3.3
320	14.0	6.0	4.7
1236	18.0	2.0	6.0
N/A	20.0	0.0	6.7

[a]The required mass of crude *s. Carlsberg* (mgCarlsberg) was calculated according to the following expression:

$$\text{mgCarlsberg} = 3\,\text{mL buffer} \times \text{EnzConc} \tag{1}$$

where EnzConc is the concentration of enzyme.
[b]The amount of fumed silica is computed with the following expressions:

$$\text{AreaEnz} = 355\,\text{m}^2/\text{g} \times (\text{mgCarlsberg}/1{,}000) \tag{2}$$

where AreaEnz is the projected area of enzyme. The specific area for *s. Carlsberg* ($355\ \text{m}^2/\text{g}$) was calculated assuming a diameter of 4.2 nm for this enzyme (24):

$$\text{Area of FS} = (\text{AreaEnz}/\%\text{SC}) \times 100 \tag{3}$$

The expression in (2) allows the calculation of the required milligrams of fumed silica as described for CALB.

5. Set the lyophilizer condenser to −50°C. Cool down the drying chamber to −40°C.
6. Transfer the preparations to the lyophilizer.
7. Evacuate the drying chamber to 160–200 mTorr (depending on the drying state).
8. Set the shelf temperature to 25°C and perform primary drying for about 48 h at this setting. The subsequent secondary drying requires about 24 h and removes the majority of the remaining water in the preparation (*see* **Fig. 12.1**).
9. Remove the samples from the lyophilizer and store them at –20°C in sealed vials (*see* **Note 5**).

3.2. Initial Reaction Rates and Kinetic Parameters

These studies are intended to elucidate the impact of the %nominal surface coverage on the catalytic performance of the immobilized enzyme on fumed silica. In all experiments, hexane was used as the reaction medium.

Table 12.2
Summary of the amount of fumed silica and buffer employed to prepare CALB-supported nanobiocatalysts with a variety of nominal surface coverages. In all cases, the amount of crude CALB was 5.83 mg

Target preparations (%nominal surface coverage)	Aqueous buffer (mL)	Fumed silica[a] (mg)	Enzyme concentration (mg/mL)	Amount of preparation containing 35 U (milligram of lyophilized preparation[b])
2	58.3	651.3	0.1	131.4
4	58.3	325.7	0.1	66.3
12	19.4	108.6	0.3	22.9
17	8.3	76.6	0.7	16.5
50	2.1	26.1	2.7	6.4
100	1.7	13.0	3.5	3.8
150	1.5	8.7	3.9	2.9
230	1.5	5.7	4.0	2.3
300	1.2	4.3	4.7	2.0
400	1.2	3.3	5.0	1.8
538	1.1	2.4	5.1	1.7
1250	0.9	1.0	6.3	1.4
2087	0.9	0.6	6.3	1.3
N/A	0.9	0.0	6.7	1.2
Novozym 435®	–	–	–	5.0

[a]The required mass of fumed silica was calculated according to the following expressions:

$$\text{Area of FS} = (3.32\,\text{m}^2/\%\text{SC}) \times 100 \quad (4)$$

where 3.32 m^2 is the projected area of enzyme (assuming a diameter of 6.4 nm for CALB (25)) and %SC is the targeted nominal surface coverage.

$$\text{mgFS} = (\text{Area of FS}/255\,\text{m}^2/\text{g}) \times 1000 \quad (5)$$

[b]The amount of preparation containing 35 U of activity (mg 35 U) is computed as follows:

$$\text{Units/mg preparation} = (\text{Units of weighed CALB})/(\text{mgFS} + \text{mgCALB}) \quad (6)$$

where mgFS is the milligram of fumed silica and mgCALB is the milligram of crude CALB.

$$\text{mg35U} = 35\ \text{U/units/mg preparation} \quad (7)$$

3.2.1. Subtilisin Carlsberg

The model reaction for studying *s. Carlsberg* was the transesterification of *N*-acetyl-l-phenylalanine ethyl ester (APEE) with 1-propanol. The initial reaction rates as a function of %nominal surface coverage are shown in **Fig. 12.2** (*see* **Note 6**):

1. Store as-received hexane (purity >99.9%, as-received water content about 10 ppm) and 1-propanol (purity 99%) over molecular sieves for drying for approximately 24 h before use.

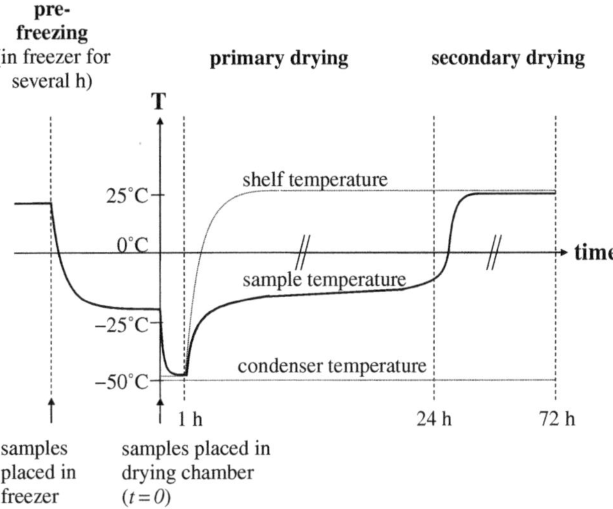

Fig. 12.1. Schematic of the temperature profiles for the sample, the drying chamber, and the condenser during the lyophilization of fumed silica/enzyme buffered suspensions.

2. Transfer 5 mL of hexane to 15-mL Teflon-lined, screw-capped glass vials.
3. Dissolve the substrates and form homogeneous solutions containing 5–40 mM APEE (purity >99%), 0.85 M of 1-propanol, and 1.5 mM nonadecane (nonreacting standard for gas chromatography analysis).
4. Add 5 mg of the lyophilized supported nanobiocatalyst to the reaction mixture.
5. Carry out the reaction at 30°C under constant shaking conditions (250 rpm) in a temperature-controlled shaker incubator.
6. Take samples of 400 μL from the homogeneous reaction mixtures in 2-mL glass vials to perform initial reaction rate measurements (*see* **Note 7**). This process is completed over a period of 30–75 min for supported nanobiocatalysts, and due to lower reaction rates for 90–270 min for fumed silica-free preparations (*see* **Note 8**).
7. Store samples at 4°C for further chromatographic analysis.

3.2.2. CALB

Two model reactions were considered for kinetic studies with CALB: the esterification of geraniol with acetic acid and the enantioselective transesterification of (*RS*)-1-phenylethanol with vinyl acetate. The initial reaction rates as a function of the %nominal surface coverage are shown in **Figs. 12.3** and **12.4**, respectively (*see* **Note 9**). The reactions were also carried out with the commercially available preparation Novozym 435®. The results

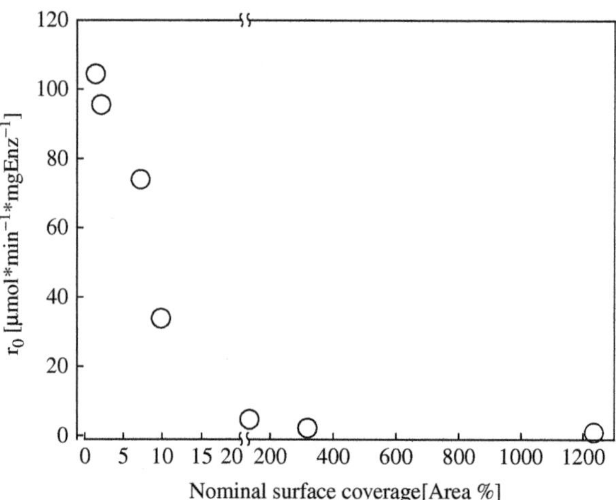

Fig. 12.2. Initial reaction rates for *s. Carlsberg* immobilized on fumed silica nanoparticles as a function of the nominal surface coverage (%SC) in the preparation. Experiments were conducted in hexane at a substrate concentration of 5 mM APEE (*see* **Note 6**).

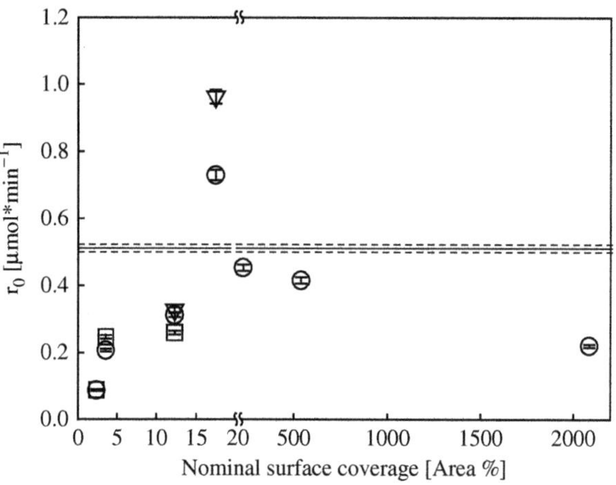

Fig. 12.3. Initial reaction rates for CALB immobilized on fumed silica nanoparticles as a function of the nominal surface coverage (%SC) in the preparation. The reactions were carried out in hexane at a substrate concentration of 0.1 M geraniol. (0) Experiments over the entire domain of surface coverages. Experiments at low surface coverages to confirm the decrease in reaction rate in this regime: (∇) first replica of 17 and 12% SC; (□) second replica of 12, 4, and 2% SC. *Solid line* represents the Novozym 435® (*see* **Table 12.3**). *Dashed lines* and *y*-error bars represent the cumulative standard error from the calculation of conversion and the linear assumptions of the initial reaction rates (*see* **Note 9**).

are superimposed for comparison in **Figs. 12.3** and **12.4** (*see* **Note 10**).

Esterification of geraniol:

1. Pipette 5 mL of hexane (as-received) to 24-mL Teflon-lined, screw-capped glass vials.

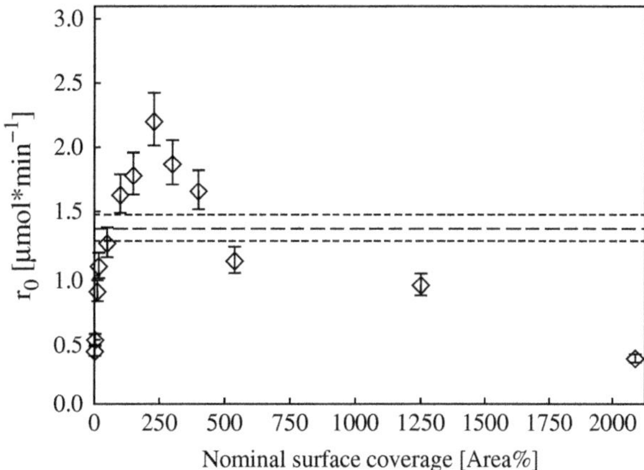

Fig. 12.4. Initial reaction rates for CALB immobilized on fumed silica nanoparticles as a function of the nominal surface coverage (%SC) in the preparation. The reactions were carried out in hexane at a substrate concentration of 82 mM (*RS*)-1-phenylethanol. *Long-dashed line* represents the Novozym 435® (*see* **Table 12.3**). *Short-dashed lines* and *y*-error bars represent the cumulative standard error from the calculation of conversion and the linear assumption of the initial reaction rates (*see* **Note 9**).

2. Add geraniol (purity 98%) to reach initial geraniol concentrations ranging from 35 to 100 mM. The initial concentration of acetic acid (purity >99.7%) was always maintained at 100 mM.
3. Add the lyophilized supported nanobiocatalyst containing 35 U of activity (*see* **Table 12.2**).
4. Carry out the reaction in the same manner as for *s. Carlsberg.*
5. Take samples of 200 μL from the homogeneous reaction mixtures in 2-mL glass vials to perform initial reaction rate measurements (*see* **Note 7**). This process is completed over a period of 65 min for supported nanobiocatalysts, and due to lower reaction rates for 120–240 min for fumed silica free preparations (*see* **Note 8**).
6. Store samples at 4°C for further chromatographic analysis.

Enantioselective transesterification of (*RS*)-1-phenylethanol:

1. Pipette 6 mL of hexane (as-received) to 12-mL glass vials with open top caps and septa.
2. Dissolve 42.4 mg of vinyl acetate (purity 99%) and 60 mg of (*RS*)-1-phenylethanol (purity 98%) (*see* **Note 11**).
3. Add the lyophilized supported nanobiocatalyst containing 35 U of activity (*see* **Table 12.2**).
4. Carry out the reaction at 30°C under constant shaking (280 rpm) in a temperature-controlled shaker incubator.

Table 12.3
Summary of apparent kinetic constants for fumed silica-supported nanobiocatalysts compared with preparations obtained with other methods

Enzyme preparation	%Nominal surface coverage (%SC)	V_{max}[c] ($\mu mol_{product} \times min^{-1} \times mg_{Enz}^{-1}$)	K_m (mM)	Catalytic efficiency (V_{max}/K_m) ($\mu mol_{product} \times min^{-1} \times mg_{Enz}^{-1} \times mM^{-1}$)
s. Carlsberg on FS (13)[a]	2	317.9	3.3	96.8
s. Carlsberg activated with 98 wt% KCl (13)	–	175.0	7.0	24.9
CALB on fumed silica (15)	17	1.8	161.0	1.14×10^{-2}
	2087	0.3	238.0	1.40×10^{-3}
Novozym 435®[b] (15)	–	1.4	204.0	6.71×10^{-3}

[a]On a wt% basis, our nanobiocatalyst with a 2% SC compares with the 98 wt% KCl+activated preparation.
[b]To compare with the commercially available Novozym 435® (CALB immobilized on macroporous polymeric resins), we take into account the fact that after immobilization on these polymeric supports, 50% of the CALB molecules initially added remain active (26).
[c]The product changes according to the reaction being analyzed. Geranyl acetate (GerAc) and *N*-acetyl-L-phenylalanine propyl ester (APPE) are the products for the reactions involving CALB and *s. Carlsberg*, respectively. All reactions involving CALB were carried out with the same mass of enzyme; therefore, the term mg_{Enz} can be factored out.

5. Sample 400 μL aliquots of the homogeneous reaction mixtures through the septum and filter them (PTFE syringe filters, pre-purged) to 2-mL glass vials for initial reaction rate measurements. This process is completed over a period of 80 min for supported nanobiocatalysts, and due to lower reaction rates for 120–240 min for fumed silica free preparations (*see* **Note 8**).
6. Store samples at 4°C for further chromatographic analysis.

3.3. Gas Chromatographic Analysis

Gas chromatographic analyses were developed to monitor the transesterification of APEE with 1-propanol and the esterification of geraniol with acetic acid.

Transesterification of APEE:

1. Transfer the stored samples to 1.5-mL Eppendorf tubes and centrifuge them for 30 s at 3,300 rpm.
2. Analyze the supernatant via gas chromatography (GC) by detecting the appearance of *N*-acetyl-L-phenylalanine propyl ester (APPE); DB-5 capillary column; He carrier gas pressure of 30 psi (1.3 mL/min), split ratio of 1/400, injection and detection at 250°C, linear column temperature ramp of 150–210°C, 8°C/min. The retention times are as follows: APEE, 5.4 min; APPE, 6.4 min; and nonadecane, 6.7 min (*see* **Note 12**).
3. Calculate the apparent kinetic constants $(V_{max})_{app}$ and $(K_m)_{app}$ by fitting the obtained initial rate data to the Michaelis–Menten equation using Lineweaver–Burk plots. The kinetic constants were obtained with respect to APEE. The catalytic efficiencies were then calculated as $(V_{max}/K_m)_{app}$ (*see* **Table 12.3**).

Esterification of geraniol:

1. Inject the stored samples to a GC to measure the formation of the esterification product geranyl acetate (GerAc); DB-WAX capillary column; He carrier at 30 psi (1.3 mL/min), split ratio of 1/200, injection and detection at 250°C, linear column temperature ramp of 56–200°C at 30°C/min. The retention times are as follows: geranyl acetate, 13.1 min; geraniol, 14.4 min (*see* **Note 12**).
2. Calculate the apparent kinetic constants as described for *s. Carlsberg*. In this case, the kinetic constants were obtained with respect to the geraniol substrate (*see* **Table 12.3**).

3.4. HPLC Analysis

The HPLC analyses were developed for the enantioselective transesterification of (*RS*)-1-phenylethanol:

1. Prepare the mobile phase for isocratic elution as a mixture of hexane/isopropyl alcohol (9:1, v/v).

2. Transfer 200 μL of the stored samples to 2-mL glass vials containing 1,100 μL of the mobile phase.
3. Inject 5 μL of the treated sample to the HPLC system to quantify the reaction products and substrates; chiracel OD-H chiral column; mobile phase at a flow rate of 0.6 mL/min, column temperature of 30°C, and PDA detector measuring absorbance at 210 nm. The retention times are as follows: vinyl acetate, 6.5 min; *R*-1-phenylethyl acetate, 7.2 min; *R*-1-phenylethanol, 10.3 min; and *S*-1-phenylethanol, 11.1 min.

4. Notes

1. Other enzymes belonging to the family of hydrolases are likely to improve their catalytic performance in nonaqueous media upon immobilization on fumed silica relative to pure enzyme.
2. The pH optimum may vary for enzymes other than those mentioned here. Adjustment to different pH values is, however, easily achievable by changing the buffer. We have demonstrated that sodium acetate and sodium bicarbonate buffer (pH 4.0 and 9.5, respectively) can be incorporated if it is required (data not shown).
3. The amount of enzyme for the studies involving CALB was maintained constant to match the units of activity present in 5 mg of the commercially available preparation – Novozym 435®. For *s. Carlsberg* the mass of enzyme was varied to achieve the proposed concentrations.
4. The %nominal surface coverage (%SC) is defined as the projected area of the enzyme molecules (assuming spherical geometry and sizes based on X-ray crystallographic analysis (21, 22)) divided by the nominal surface area of fumed silica (provided by the manufacturer).
5. Before using the preparations, the sealed vials need to be warmed to room temperature. It is also recommended to reduce as much as possible the exposure of the samples to ambient moisture.
6. **Figure 12.2** shows that the reaction rate increases by two orders of magnitude as the nominal surface coverage approaches 2% SC. This low surface coverage regime is, therefore, recommended to prepare highly active fumed silica-based nanobiocatalysts involving *s. Carlsberg*.

7. Initial reaction rates were determined from linear fits over the average values of the GC measurements taken in duplicate.
8. The times for enzymatic assays may vary according to the available active sites in the sample. Linearity in the initial reaction rates must be assured. For the reactions under consideration, this condition is usually attained at conversions below 20%.
9. **Figures 12.3** and **12.4** show independently that for CALB as opposed to *s. Carlsberg*, an intermediate nominal surface coverage is required to prepare nanobiocatalysts with exceptional activity. Above and below this coverage, issues regarding either mass transfer limitations or conformational changes are thought to be detrimental for the catalytic competency of the enzyme. At high nominal surface coverages, the enzyme molecules are likely to be agglomerated leading to a diffusional resistance for the substrate molecules to reach the active sites (20). At low surface coverages, however, the enzyme molecules are likely to be separated from each other. This may lead to increased surface–protein interactions that are likely to promote substantial structural changes. This behavior has been associated with enzymes with high structural flexibility (11, 12, 15, 20, 23).
10. Commercial Novozym 435® was stored at 4°C and used as-received. Novozym 435® is reported to be CALB immobilized on macroporous acrylic particles (0.3–0.9 mm diameter) with a reported catalytic activity of about 7,000 PLU/g (propyl laurate units per gram).
11. These amounts correspond to an equimolar concentration of 82 mM for the substrates involved.
12. In all GC analyses, 1 μL of the withdrawn sample was injected.

References

1. Gross, R. A., Kumar, A., and Kalra, B. (2001) Polymer synthesis by in vitro enzyme catalysis. *Chem. Rev.* **101**, 2097–2124.
2. Gross, R. A., and Kalra, B. (2002) Biodegradable polymers for the environment. *Science* **297**, 803–807.
3. Barahona, D., Pfromm, P. H., and Rezac, M. E. (2006) Effect of water activity on the lipase catalyzed esterification of geraniol in ionic liquid [bmim]PF6. *Biotechnol. Bioeng.* **93**, 318–324.
4. Bartling, K., Thompson, J. U. S., Pfromm, P. H., Czermak, P., and Rezac, M. E. (2001) Lipase-catalyzed synthesis of geranyl acetate in *n*-hexane with membrane-mediated water removal. *Biotechnol. Bioeng.* **75**, 676–681.
5. Gotor, V. (2002) Biocatalysis applied to the preparation of pharmaceuticals. *Org. Process Res. Dev.* **6**, 420–426.
6. Klibanov, A. M. (2001) Improving enzymes by using them in organic solvents. *Nature* **409**, 241–246.

7. Ghanem, A. (2007) Trends in lipase-catalyzed asymmetric access to enantiomerically pure/enriched compounds. *Tetrahedron* **63**, 1721–1754.
8. Klibanov, A. M. (1997) Why are enzymes less active in organic solvents than in water? *Trends Biotechnol.* **15**, 97–101.
9. Persson, M., Wehtje, E., and Adlercreutz, P. (2002) Factors governing the activity of lyophilised and immobilised lipase preparations in organic solvents. *ChemBioChem* **3**, 566–571.
10. Long, J., Hutcheon, G. A., and Cooper, A. I. (2007) Combinatorial discovery of reusable noncovalent supports for enzyme immobilization and nonaqueous catalysis. *J. Comb. Chem.* **9**, 399–406.
11. Chen, B., Miller, E. M., Miller, L., Maikner, J. J., and Gross, R. A. (2007) Effects of macroporous resin size on Candida antarctica lipase B adsorption, fraction of active molecules, and catalytic activity for polyester synthesis. *Langmuir* **23**, 1381–1387.
12. Chen, B., Miller, M. E., and Gross, R. A. (2007) Effects of porous polystyrene resin parameters on Candida antarctica lipase B adsorption, distribution, and polyester synthesis activity. *Langmuir* **23**, 6467–6474.
13. Wurges, K., Pfromm, P. H., Rezac, M. E., and Czermak, P. (2005) Activation of subtilisin Carlsberg in hexane by lyophilization in the presence of fumed silica. *J. Mol. Catal. B Enzym.* **34**, 18–24.
14. Pfromm, P. H., Rezac, M. E., Wurges, K., and Czermak, P. (2007) Fumed silica activated subtilisin Carlsberg in hexane in a packed-bed reactor. *AIChE J.* **53**, 237–242.
15. Cruz, J. C., Pfromm, P. H., and Rezac, M. E. (2009) Immobilization of *Candida antarctica* lipase B on fumed silica. *Process Biochem.* **44**, 62–69.
16. Gun'ko, V. M., Mironyuk, I. F., Zarko, V. I., Voronin, E. F., Turov, V. V., Pakhlov, E. M., Goncharuk, E. V., Nychiporuk, Y. M., Vlasova, N. N., Gorbik, P. P., Mishchuk, O. A., Mishchuk, O. A., Chuiko, A. A., Kulik, T. V., Palyanytsya, B. B., Pakhovchishin, S. V., Skubiszewska-Zieba, J., Janusz, W., Turov, A. V., and Leboda, R. (2005) Morphology and surface properties of fumed silicas. *J. Colloid Interface Sci.* **289**, 427–445.
17. Voronin, E. F., Gun'ko, V. M., Guzenko, N. V., Pakhlov, E. M., Nosach, L. V., Leboda, R., Skubiszewska-Zieba, J., Malysheva, M. L., Borysenko, M. V., and Chuiko, A. A. (2004) Interaction of poly(ethylene oxide) with fumed silica. *J. Colloid Interface Sci.* **279**, 326–340.
18. Gun'ko, V. M., Voronin, E. F., Nosach, L. V., Pakhlov, E. M., Guzenko, N. V., Leboda, R., and Skubiszewska-Zieba, J. (2006) Adsorption and migration of poly(vinyl pyrrolidone) at a fumed silica surface. *Adsorption Sci. Technol.* **24**, 143–157.
19. Gun'ko, V. M., Mikhailova, I. V., Zarko, V. I., Gerashchenko, I. I., Guzenko, N. V., Janusz, W., Leboda, R., and Chibowski, S. (2003) Study of interaction of proteins with fumed silica in aqueous suspensions by adsorption and photon correlation spectroscopy methods. *J. Colloid Interface Sci.* **260**, 56–69.
20. Bosley, J. A., and Peilow, A. D. (1997) Immobilization of lipases on porous polypropylene: Reduction in esterification efficiency at low loading. *J. Am. Oil Chem. Soc.* **74**, 107–111.
21. Neidhart, D. J., and Petsko, G. A. (1988) The refined crystal-structure of subtilisin Carlsberg at 2.5 A resolution. *Protein Eng.* **2**, 271–276.
22. Uppenberg, J., Hansen, M. T., Patkar, S., and Jones, T. A. (1994) Sequence, crystal-structure determination and refinement of 2 crystal forms of lipase-B from *Candida antarctica*. *Structure* **2**, 293–308.
23. Koutsopoulos, S., van der Oost, J., and Norde, W. (2005) Structural features of a hyperthermostable endo-beta-1, 3-glucanase in solution and adsorbed on "invisible" particles. *Biophys. J.* **88**, 467–474.
24. Shaw, A. K., and Pal, S. K. (2007) Activity of subtilisin Carlsberg in macromolecular crowding. *J. Photochem. Photobiol. B Biol.* **86**, 199–206.
25. Sate, D., Janssen, M. H. A., Stephens, G., Sheldon, R. A., Seddon, K. R., and Lu, J. R. (2007) Enzyme aggregation in ionic liquids studied by dynamic light scattering and small angle neutron scattering. *Green Chem.* **9**, 859–867.
26. Chen, B., Hu, J., Miller, E. M., Xie, W. C., Cai, M. M., and Gross, R. A. (2008) Candida antarctica lipase B chemically immobilized on epoxy-activated micro- and nanobeads: Catalysts for polyester synthesis. *Biomacromolecules* **9**, 463–471.

Chapter 13

Microencapsulation of Bioactive Nanoparticles

Fei Gao, Ping Wang, and Guanghui Ma

Abstract

Supported or modified enzymes in the form of mobile nanoparticles are designed for enhanced activities and stabilities; however, their practical operations are dwarfed due to their tiny size which always makes recycling an arduous task and a potential risk to the environment. To overcome such drawbacks, this chapter describes a method for the preparation of a new form of microcapsules, possessing single-cavity compartments and nano-pores in the shell, to encage nanoparticle-based biocatalysts and form cell-like microreactors (CLMRs). The encaged nanoscale catalysts are maintained their high activities as in a bulk-phase solution, while they could be handled as materials of sizes hundreds-fold larger.

Key words: Encapsulation, enzyme immobilization, double emulsion, hierarchical structure, nanoparticles.

1. Introduction

Enzymes are excellent but delicate natural catalysts, calling for appropriate immobilization approach to enhance their stability and facilitate their applications. Nanodispersed carriers such as nanoparticle (1, 2), nanocontainer (3), and single-molecule armor (4) have been intensively studied as supports for enzymes and were believed to accommodate the usual contradictory requirements in practice: minimum diffusion restriction and maximum loading capacity (3, 5). What is more fascinating, the mobility of the nanoscale carriers in solution greatly contributes to their performance (1), as is featured by Brownian motion, and is thus altogether distinct from heterogeneous catalysis. Such nanoparticle-based biocatalysts (NBBCs) are gifted in action, however, they would be dwarfed in practical application, for their tiny size makes recycling an arduous task and a potential risk

P. Wang (ed.), *Nanoscale Biocatalysis*, Methods in Molecular Biology 743,
DOI 10.1007/978-1-61779-132-1_13,

to the environment. Nanodispersed engineered materials tend to suspend in aqueous medium or in air for a long time, and therefore can hardly be digested spontaneously. Although it is still hard to assess the environmental impacts, protective research is necessary to ensure a sustainable development of the nanotechnology (6). This study aimed to develop a hierarchically ordered structure to fulfill practical applications of nanodispersed functional materials and to suppress environmental risks.

To overcome the recycling difficulty and to guarantee application security, NBBCs first constructed are managed to assemble as cell-like microreactors (CLMRs), using porous shells as boundaries. The original mobility and activity of nanodispersed biocatalysts are well preserved in the hollow microcapsules, though the size of the assembled biocatalysts, now tens of micrometers, will involve little difficulty in the control. Compared to other hierarchical approaches, e.g., hosting NBBCs in mesoporous material (7), or embedding enzyme-laden particles in gel beads (8), a remarkable asset of the proposed strategy is that the assembled nanoparticles are still enjoying their original mobility, in the form of Brownian motion, which essentially guarantees the native-like performance.

Quite a number of elegant methods have been developed to prepare polymer microcapsules suitable for a variety of core materials. However, it is still hard to find a strategy to compromise the mechanical strength, capability, permeability, and biocompatibility by a simple procedure. Traditionally in polymeric field, hollow microspheres and microcapsules could be produced by phase separation between polymers and hydrophobic solvent, during either polymerization or solvent evaporation from the polymer solution droplets (9–14). The polymer forms a shell and the solvent is engulfed to be a liquid core, which could be removed by subsequent extraction. Some other researches used existing spheres as seeds or raw materials, internal compartments of which could be fabricated by thermal or chemical treatment (8, 11, 15). Such kinds of procedures could be well explained by thermodynamic principles and would lead to an easy and accurate control. However, applications, especially encapsulations of biomaterial, are often restricted by the hydrophobic cores and the harsh procedure. In order to meet the requirements from delicate biomaterial during some practical encapsulation, an aqueous core is usually desired for core materials, such as calcium alginate capsules. Although gel-based capsules can be prepared through a simple and easy procedure in water phase, which offers a sustainable biocompatible environment, they usually suffer extensively from soft texture and poor stability (16).

Another well-known approach to entrap water solution is the two-step emulsification, to form water-in-oil-in-water (W/O/W) double emulsions, which have been studied ever since as early as 1925 (17). Combining with subsequent solidification, such as suspension polymerization or solvent evaporation (18–20), such liquid oil globules could be transformed to solid microcapsules. For the hollow structure and unique release behavior, a variety of applications are showing interest in such W/O/W microcapsule (21). Although a rather large literature exists, there is still a lack of general formula or quantitative criteria on the multiple emulsion (22), and the preparation condition was often determined empirically and varied from case to case. Generally, to prepare W/O/W emulsion, the primary emulsification should be conducted more intensively than the subsequent one, leading to oil globules containing quite a number of water droplets inside. It was the reason why W/O/W globules were usually endowed with multiple-cell (or called multi-compartment) morphology, and so were the double-emulsion templated microcapsules. Such commonly obtained morphology was believed to be the most stable and an ideal structure for release control (23); however, it would be short in volume of each single cavity and would slow down the mass transportation across the capsule shells.

In this chapter, we present a novel type of hollow microcapsule to encage the pre-constructed NBBCs, taking advantage of the bio-friendly procedure of W/O/W emulsification and the rigid/porous capsule structure obtained from suspension polymerization. We also present an effective strategy to manipulate the morphology of the W/O/W emulsion globules, to form single-core water-in-oil (W/O) globules based on their multi-core precursors. Previously, such single-core liquid microcapsules could only be fabricated one by one in micro-fluidic devices (24). On the basis of single-core W/O/W emulsion globules, the thickness of their oil membranes could be easily controlled by osmotic gradient between W_1 and W_2. The eventual polymerization is initiated by ultraviolet radiation with a wavelength of 380 nm, which has also proved to be a compatible process for biomolecules (25). In addition, phase-separation mechanism is employed to produce nano-channels across the shells. Short-chain alcohols play a primary part in the mixture solvents (or called porogen), the mechanism of which has been well studied and widely accepted (26–30). Besides acrylate monomer, cross-linkers which possess more than one vinyl group of each molecule should be essentially employed to form porous matrix and rigid structure. Channels across the shells of the CLMR can be precisely designed to be suitable only for rapid exchange of substrates and products in the form of small molecules and not to allow the NBBCs to escape. Such hierarchical strategy can well accommodate another pair of contradictions in enzyme supporting: maximal stability for assembled cores and minimal resistance for molecule diffusion, as it is more comprehensive and practical than direct encapsulation.

2. Materials

2.1. Soap-Free Emulsion Polymerization

1. Oil phase for emulsion polymerization includes both methyl methacrylate (MMA) as the primary monomer and methacrylic acid (MAA) as the hydrophilic monomer, with a molar ratio at 20:1.
2. Potassium persulfate ($K_2S_2O_8$) is dissolved in deionized water as initiator.
3. Water should be deionized using ion-exchange resins.

2.2. Seeded Polymerization

1. Additional oil phase includes carboxyl monomer methacrylic acid (MAA), crosslinker ethylene dimethacrylate (EDMA, Sigma), and hexanol (HA) which is important for the dispersity of the latex during the followed polymerization.
2. 2,2′-Azobis(2,4-dimethylvaleronitrile) (ADVN) is dissolved in the additional oil phase as initiator.

2.3. Covalent Coupling for Enzyme

1. Water-soluble carbodiimide, 1-ethyl-3-(3-dimethylamino-propyl carbodiimide), is used to activate the carboxyl groups on nanoparticles, to couple with the amino groups exposing on enzyme molecules.
2. Appropriate buffer solution with specific pH value is desired by given immobilization with specific enzyme. For example, a phosphate buffer with a pH value at 6.5 and concentration at 100 mM is desired during the covalent coupling for a α-chymotrypsin, and a Tris–HCl buffer with a pH at 7.8 and concentration at 80 mM is desired during the followed separation and storage.
3. Solid concentration of the primary carrier is adjusted to about 10% in deionized water.

2.4. Double Emulsion and Suspension Polymerization

1. Acrylate vinyl monomer and cross-linkers are chosen as the primary components of oil phase. MMA, EDMA, and trimethylolpropane trimethacrylate (TTMA, Sigma) should be distilled under reduced pressure to remove inhibitors before utilization.
2. Oil-soluble initiator ADVN is employed to initiate the suspension polymerization, which can be initiated either by thermal treatment or by ultraviolet irradiation. The molar ratio between initiator molecule and vinyl group in the oil is 1:500.
3. Mid-chain alcohol such as hexanol plays the main role in mixture porogen in oil phase, which also include chloroform to mediate the density of the mixture solvent.

4. Span 85 is employed as the oil-soluble surfactant and occupies as high as 10% of the oil phase.
5. W_1 includes polymer stabilizer poly(vinyl alcohol) (PVA 217, degree of polymerization 1660, degree of hydrolysis 88.5%) to enhance the stability of embedded droplets. The concentration of PVA in W_1 ranges from 0.5 to 1.5%. Polyoxyethylene sorbitan monolaurate (Tween 20) is also desired in W_1, helping to form finer water droplets during the primary emulsification. Inorganic salt such as NaCl is necessary in W_1, helping to maintain a desired osmotic pressure. The concentration of NaCl in W_1 ranges from 0.2 to 1.0%.
6. Original part of W_2 (W_{2-1}) contains similar solutes as W_1. The PVA concentration ranges from 1.5 to 2.5%, obviously higher than in W_1. The NaCl concentration in W_2 ranges from 0.2 to 0.5%, to balance the osmotic gradient across the oil membrane during the phase second step emulsification. Sodium dodecyl sulfate (SDS) is added as water-soluble surfactant.
7. Additional part of W_2 (W_{2-2}) which can be several times larger than the original volume is employed to further adjust the original solute concentration after the secondary emulsification. The PVA concentration ranges from 2.5 to 4.0%, helping to enhance the external viscosity and the stability of external interfaces of double emulsion. Surfactant and salt are not necessary in W_{2-2}.

2.5. Collection and Storage for CLMRs

1. Washing solution contains 20% ethanol, which is used to remove the residual organic solvent and oil-soluble surfactant embedding in the porous shells after the suspension polymerization.
2. Appropriate buffer solution is used to store enzyme-loaded microcapsules. For example, when α-chymotrypsin is loaded, 0.08 M Tris–HCl (pH 7.8) is desired.

3. Methods

Two groups of P(MMA–EDMA–MAA) nanoparticles, 166 and 390 nm in diameters (surface weighted mean, Span<0.60), respectively, are employed as the primary carriers for enzyme molecules. A model enzyme α-CT is first immobilized on the surface of the polymer nanoparticles by covalent bonding. The side carboxyl group of the co-monomer MAA provides the points for covalent binding. The nanoparticle-based biocatalysts are further encaged within cell-like microcapsules. These microcapsules are prepared by a two-step emulsification, emulsion ripening and suspension polymerization.

The original mobility and activity of nanodispersed biocatalysts would be well preserved in the hollow microcapsules, though the size of the assembled biocatalysts, now tens of micrometers, would involve little difficulty in the control. Such hierarchical scaling-up strategy could accommodate another pair of contradictions in enzyme supporting, maximal stability for assembled cores and minimal resistance for molecule diffusion, as it is more comprehensive and practical than direct encapsulation for enzymes.

3.1. Preparation for Nanoparticles

1. Seed nanoparticles are synthesized during the first stage of soap-free emulsion polymerization, which is carried out in a four-neck flask equipped with an anchor agitator and a nitrogen inlet nozzle.
2. Aqueous phase is bubbled with nitrogen for an hour before the subsequent polymerization, which should also be carried out under nitrogen atmosphere all through.
3. To prepare 166 nm seeds, an mixture of 16 g MMA and 0.8 g MAA is emulsified into 1,600 mL water phase by mechanical stirrer. To prepare 399 nm seeds, the same oil mixture is emulsified into 200 mL water phase.
4. Agitate the emulsion at 400 rpm and heat it to 70°C by water bath.
5. Add 0.24 g K_2SO_4 to the aqueous solution for 166 nm seeds and 0.15 g for 390 nm seeds. Incubate the emulsion at 70°C for 2 h, during which seed nanoparticles are formed. Then cool the latex to room temperature.
6. Add a fresh oil phase with 2 g MAA, 2 g EDMA, 2 g HA, and 50 mg ADVN to the seed latex and stir the mixture at 400 rpm. The seed nanoparticles will absorb the additional monomers and crosslinkers in 10 h.
7. A secondary polymerization is initiated by heating the mixture to 70°C and lasts for 6 h to convert all monomers involved.
8. The prepared latex is first dialysed with 2–3 L 20% ethanol solution three times to remove HA and ion, and then with 2–3 L deionized water for at least five times to remove the residual ethanol. **Figure 13.1** shows the uniform nanoparticles of two different sizes.
9. The solid concentration of final latex can be further adjusted by evaporation.

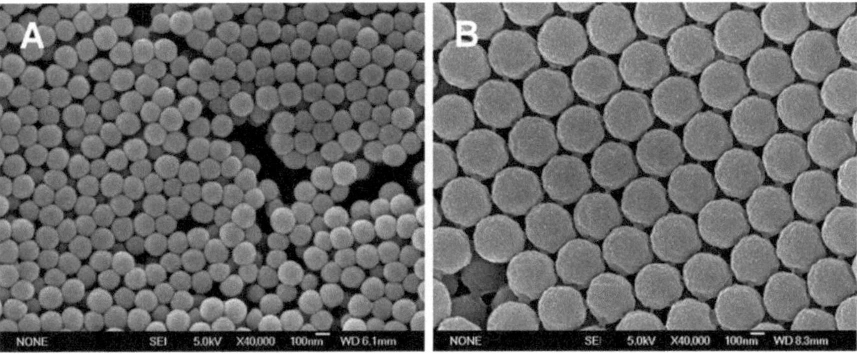

Fig. 13.1. Scanning electronic microscope (SEM) images of P(MMA–EDMA–MAA) nanoparticles: (**a**) 166 nm particles and (**b**) 390 nm particles.

3.2. Enzyme Supporting by Nanoparticles

1. Prepare enzyme solution by dissolving α-CT into 1 mL 100 mM phosphate buffer, pH 6.5, to obtain a final concentration at 4 mg/mL for further reaction with 166 nm particles and at 2 mg/mL with 390 nm particles.
2. Prepare activator solutions by dissolving EDC in deionized water, at 30 mM for coupling with 166 nm nanoparticles and 10 mM for the 390 nm nanoparticles.
3. Mix the latex of nanoparticles with a phosphate buffer pH 6.5, to form working latex consisting of 3% solid and 0.1 M phosphate.
4. Activate the carboxyl groups exposing on nanoparticles by adding 1 mL EDC solution into 3 mL working latex. The activation is carried out in a sealed bottle, shaken by a shaking table with a rotating speed at 150 rpm for 10 min. The chemical route for coupling is shown in **Fig. 13.2**.
5. Add 1 mL enzyme solution to the latex of activated nanoparticles and shake the latex on shaking table with a rotating speed at 100 rpm for another 2 h.

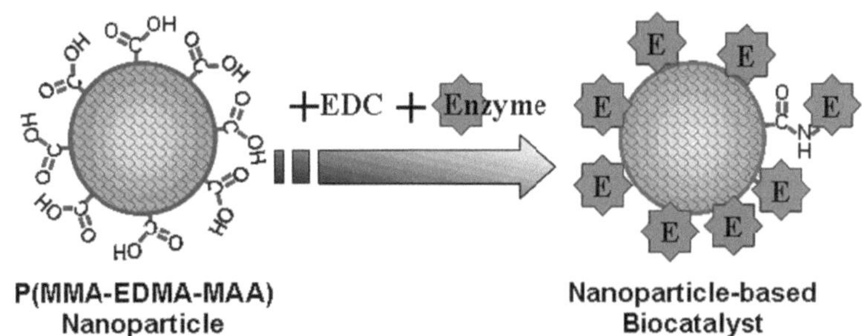

Fig. 13.2. Chemical route for NBBCs construction.

6. The freshly obtained NBBCs are then washed several times with fresh Tris–HCl buffer, by centrifugation, till no protein can be detected in the supernatant.
7. The NBBCs are finally suspended in 80 mM Tris–HCl buffer solution, pH 7.8, and stored at 4°C.

3.3. Encapsulation of Bioactive Nanoparticles

1. Primarily, add the latex of NBBCs into W_1 to form NBBCs/W_1, with a solid concentration ranging from 0.1 to 10%.
2. Emulsify a certain amount of NBBC/W_1 into the oil phase with a mixing ratio at 1/10(v/v), to form a primary emulsion NBBC/W_1/O. The first step of emulsification is conducted by a homogenizer at 6,000 rpm for 2 min. The dispersed phase is emulsified to fine droplets, less than 1 μm in diameter.
3. Emulsify the primary emulsion NBBC/W_1/O into the initial part of external aqueous phase W_{2-1} with a mixing ratio at 1/10(v/v), to form a multiple emulsion NBBC/W_1/O/W_{2-1}. The secondary emulsification is conducted by mechanical stirring at 390 rpm for 3 min, producing relatively larger oil globules, tens of micrometers in diameter.
4. Add W_{2-2} to the external aqueous W_{2-1} after the two-step of emulsification, to dilute the salt concentration and increase the PVA concentration in external phase.
5. Load the emulsion into several quartz bottles, with each bottle being halfly filled with emulsion and further filled up with nitrogen. Shake these bottles with a vertical oscillator at 10 rpm, to prevent creaming or depositing of the emulsion globules.
6. The morphology evolution of these emulsion globules from multi-core to single core, which has been named as emulsion ripening, is schematically illustrated in **Fig. 13.3**. It takes an hour or more to get the maximal proportion of the single-cell microcapsules under mild rotation.
7. Put the quartz bottles directly under UV irradiation (380 nm in wavelength), to start the suspension

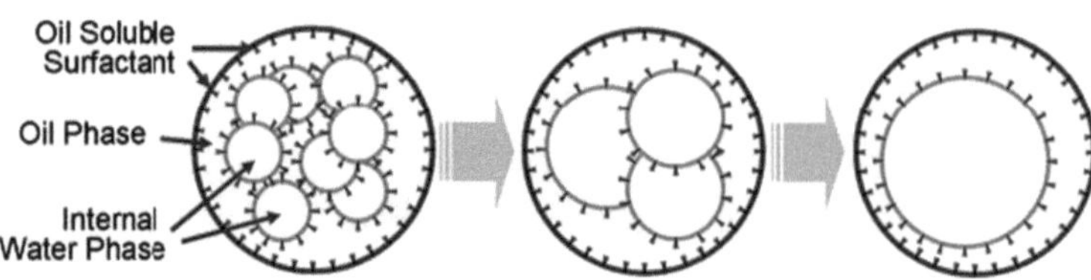

Fig. 13.3. Schematic representation of ripening of W_1/O globules.

polymerization, with all reaction bottles being shaken with a rotation speed at 60–100 rpm.

8. The O layer will solidify in a few minutes, forming solid CLMRs. The conversion of organic monomers will exceed 99% after 6 h polymerization. By choosing different ripening times, we will obtain microcapsules with different morphology. The evolution tendency of capsule morphology can be illustrated by a triangle diagram as illustrated in **Fig. 13.4**.
9. Collect and wash the CLMRs by a decompress filter. The CLMRs should be washed with warm deionized water (40°C) three times at first to remove the residual PVA adsorbing on the shells, then with 20% ethanol solution three times to remove the organic residual embedding in the shell, and with 80 mM Tris–HCl buffer solution (pH 7.8) at last to maintain a bio-friendly microenvironment.
10. Store the CLMRs in an appropriate buffer solution, such as 80 mM Tris–HCl buffer pH 7.8 at 4°C, to protect the incorporated biocatalysts. The CLMRs can be applied

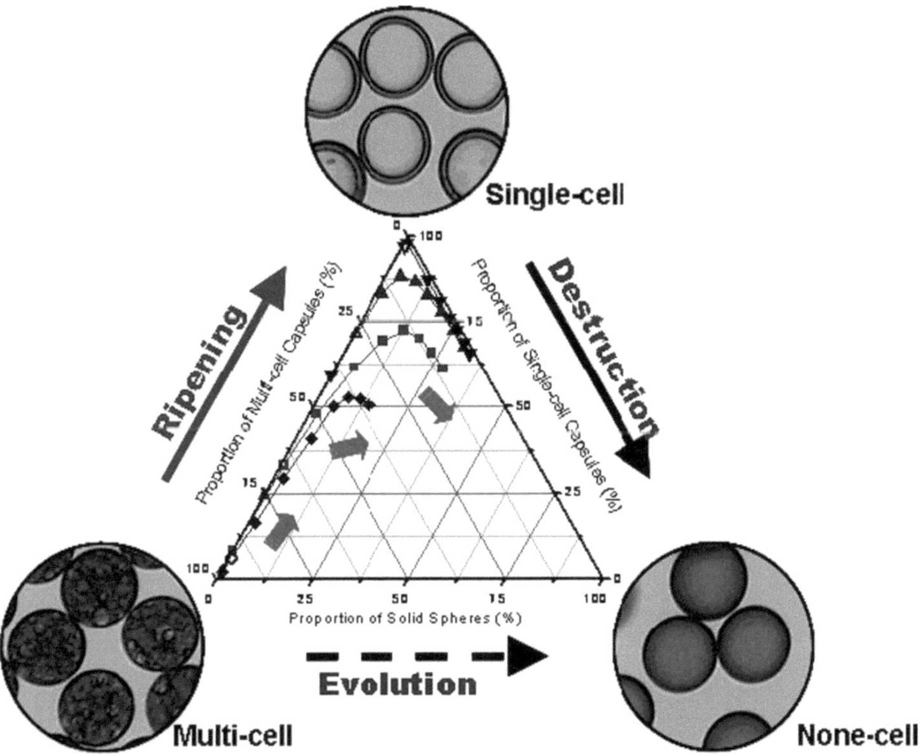

Fig. 13.4. Example ternary proportion diagram exhibiting evolution profiles of $W_1/O/W_2$ emulsions. Each *dot* stands for a unique morphology distribution of the three types with a certain recipe and ripening time. Example images attached aside exhibit the exclusive morphology of three different forms.

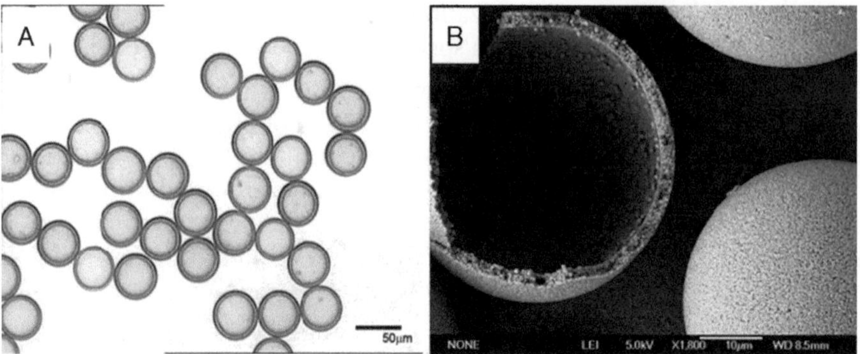

Fig. 13.5. Optical and SEM images of CLMRs: (**a**) 40–50 m capsules in aqueous solution and (**b**) NBBCs assembled in capsules.

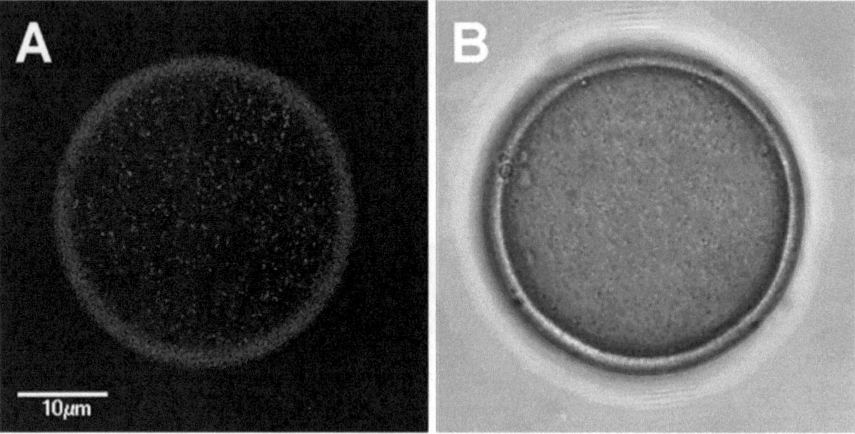

Fig. 13.6. Laser confocal scanning microscope images of a capsule containing latex of NBBCs (390 nm) dyed with rhodamine B: (**a**) fluorescent image and (**b**) bright-field image.

either in packed bed reactor or in suspension reactor. Examples of the final images of CLMRs are shown in **Figs. 13.5** and **13.6**.

4. Notes

1. Nanoparticles which are polymerized by one-step soap-free polymerization are not suitable for further encapsulation by two-step emulsification. For such nanoparticles prefer to adhere to the water/oil interfaces and can be dissolved by oil phase. Seeded polymerization is essential to strengthen the particle structure and protect the hydrophilic surfaces.

2. During the seeded polymerization, mid-chain alcohol, such as hexanol, is added to guarantee the dispersity of nanoparticles during the second step of polymerization. Although the mechanism is not clear, mid-chain alcohol is proved as an indispensable oil component when both carboxyl monomer and vinyl crosslinker are altogether involved.
3. Compared to other polymer nanoparticles reported as enzyme carriers, these particles presented possess not only available surface for enzyme immobilization but also cross-linked texture to meet further encapsulation.
4. During the encapsulation, density of the core phase W_1 has been measured at 1.010–1.020 g/cm^3. Density of the oil phase should be adjusted to the same level to maintain a relative long-term stability of the $W_1/O/W_2$. High-density component, such as chloroform, is employed as the density modifier.
5. Oil-soluble polymer, such as linear PMMA, can be used as viscocity modifier of the oil phase. Stability of W/O/W emulsion will be enhanced obviously with 1% PMMA incorporated in oil, though the emulsion ripening will be slowed down, leading to a majority proportion of multi-cell morphologies.
6. Ternary diagram is employed to exhibit the profiles of emulsion evolutions, and an example diagram is illustrated in **Fig. 13.4**. Respectively, each peak of the triangle diagram stands for an exclusive morphologies distribution, as illustrated by example images attached. The morphology distribution with a certain ripening could be represented by a single dot on the diagram, and curves connecting these dots show the evolution tendency of W/O/W emulsion. The single-cell microcapsules being the target products, the sharper curve with the higher single-cell proportion is expected.
7. Surfactants, stabilizer, and salt are essential components for both the W_1 and the W_2, the recipes of which are of guiding significance to subsequent emulsion ripening and would encode the morphology information of the final microcapsules.
8. During the microcapsule preparation, the PVA concentration in W_2 is the primary factor to affect the evolution process of the $W_1/O/W_2$ emulsion, the more PVA dissolved in W_2 will lead to the better stabilized W_1/O globules.
9. PVA also plays an important role in the internal aqueous phase, while the highest single-cell proportion will only be obtained with a mild concentration, neither too low nor too high PVA concentration in W_1.

10. Small amount of Tween 20 or Tween 80 in W_1 helps to fabricate fine aqueous droplets during the primary emulsification, leading to a better stabilized W/O emulsion, while other ionic surfactants in W_1, such as SDS, will accelerate the collapsing of double emulsion.
11. Usually, the salt concentration in W_1 should be higher than that in W_2 to promote emulsion ripening and fabricate thin oil membrane. To promote ripening, salt concentration in W_1 should be four times higher than that in W_2 at least.
12. During the secondary emulsification, salt concentration in W_{2-1} should be as high as the concentration in W_1. After the addition of W_{2-2}, salt concentration in W_2 is diluted and is obviously lower than W_1. Water molecules will diffuse inside to swell the trapped aqueous solution and to promote W_1 aggregate.
13. The first step of emulsification, which is conducted by intensive shearing, causes the majority denaturation of biocatalysts during the whole procedure. Practical operation should compromise between even encapsulation and activity maintaining, to stir as mild as possible.
14. The suspension polymerization is initiated by ultraviolet irradiation (380 nm in wavelength), using ADVN as initiator. Such mild UV irradiation is critical to guarantee the bioactivity of assembled biocatalysts.
15. The UV lights should be settled on the top of the emulsion during the UV-initiated polymerization, in order to get a even shell thickness of each microcapsule.
16. Phase-separation mechanism is employed during the polymerization to fabricate nanoscale channels across the shells, with mid-chain alcohol, such as HA, as the primary porogen. By adjusting the amounts of porogen and cross-linkers, pore sizes are prepared from 18 to 40 nm, with overall porosities from 20 to 45%.

Acknowledgments

The authors thank the National Basic Research Program of China (New 973 Program, Contract Nos. 2009CB724705) and Chinese National Science Foundation of China (20576135, 20536050, and 20728607) for their support. The support from Chinese Academy of Sciences for international collaboration is also greatly appreciated.

References

1. Jia, H., Zhu, G., and Wang, P. (2003) Catalytic behaviors of enzymes attached to nanoparticles: The effect of particle mobility. *Biotechnol. Bioeng.* **84**, 406–414.
2. Kang, K., Kan, C., Du, Y., and Liu, D. (2005) Synthesis and properties of soap-free poly(methyl methacrylate-ethyl acrylate-methacrylic acid) latex particles prepared by seeded emulsion polymerization. *Eur. Polym. J.* **41**, 439–445.
3. Wang, P. (2006) Nanoscale biocatalyst systems. *Curr. Opin. Biotechnol.* **17**, 574–579.
4. Kim, J., and Grate, J. W. (2003) Single-enzyme nanoparticles armored by a nanometer-scale organic/inorganic network. *Nano Lett.* **3**, 1219–1222.
5. Kim, J., Grate, J. W., and Wang, P. (2006) Nanostructures for enzyme stabilization. *Chem. Eng. Sci.* **61**, 1017–1026.
6. Colvin, V. L. (2003) The potential environmental impact of engineered nanomaterials. *Nat. Biotechnol.* **21**, 1166–1170.
7. Kim, J., Jia, H., Lee, C.-W., Chung, S.-W., Kwak, J. H., Shin, Y., Dohnalkova, A., Kim, B.-G., Wang, P., and Grate, J. W. (2006) Single enzyme nanoparticles in nanoporous silica: A hierarchical approach to enzyme stabilization and immobilization. *Enzyme Microb. Technol.* **39**, 474–480.
8. Fan, C.-H., and Lee, C.-K. (2001) Purification of D-hydantoinase from adzuki bean and its immobilization for N-carbamoyl-D-phenylglycine production. *Biochem. Eng. J.* **8**, 157–164.
9. Ma, G.-H., Su, Z.-G., Omi, S., Sundberg, D., and Stubbs, J. (2003) Microencapsulation of oil with poly (styrene-N,N-dimethylaminoethylaminoethyl methacrylate) by SPG emulsification technique. *J. Colloid Interface Sci.* **266**, 282–294.
10. Ma, G.-H., Chen, A.-Y., Su, Z.-G., and Omi, S. (2003) Preparation of uniform hollow polystyrene particles with large voids by a glass-membrane emulsification technique and a subsequent suspension polymerization. *J. Appl. Polym. Sci.* **87**, 244–251.
11. Okubo, M., and Minami, H. (1996) Control of hollow size of micron-sized monodispersed polymer particles having a hollow structure. *Colloid Polym. Sci.* **274**, 433–438.
12. Ma, G.-H., Omi, S., Dimonie, V. L., Sudol, E. D., and El-Aasser, M. S. (2001) Study of the preparation and mechanism of formation of hollow monodisperse polystyrene microspheres by SPG (Shirasu Porous Glass) emulsification technique. *J. Appl. Polym. Sci.* **85**, 1530–1543.
13. Okubo, M., Konishi, Y., and Minami, H. (2004) Production of hollow particles by suspension polymerization of divinylbenzene with nonsolvent. *Progr. Colloid Polym. Sci.* **124**, 54–59.
14. Okubo, M., Konishi, Y., Inohara, T., and Minami, H. (2002) Production of hollow polymer particles by suspension polymerizations for ethylene glycol dimethacrylate/toluene droplets dissolving styrene-methyl methacrylate copolymers. *J. Appl. Polym. Sci.* **86**, 1087–1091.
15. Im, S. H., Jeong, U., and Xia, Y. (2005) Polymer hollow particles with controllable holes in their surfaces. *Nat. Mater.* **4**, 671–675.
16. George, M., and Abraham, T. E. (2006) Polyionic hydrocolloids for the intestinal delivery of protein drugs: Alginate and chitosan – A review. *J. Control. Release* **114**, 1–14.
17. Seifriz, W. (1925) Studies in emulsions. III–V. *J. Phys. Chem.* **29**, 738–749.
18. Zydowicz, N., Nzimba-Ganyanad, E., and Zydowicz, N. (2002) *Polym. Bull.* **47**, 457–463.
19. Liu, R., Ma, G., Meng, F.-T., and Su, Z.-G. (2005) PMMA microcapsules containing water-soluble dyes obtained by double emulsion/solvent evaporation technique. *J. Control. Release* **103**, 31–43.
20. Meng, F. T., Ma, G. H., Qiu, W., and Su, Z. G. (2003) W/O/W double emulsion technique using ethyl acetate as organic solvent: Effects of its diffusion rate on the characteristics of microparticles. *J. Control. Release* **91**, 407–416.
21. Engel, R. H., Riggi, S. J., and Fahrenbach, M. J. (1968) Insulin: Intestinal absorption as water-in-oil-in-water emulsions. *Nature* **219**, 856–857.
22. Ficheux, M.-F., Bonakdar, L., Leal-Calderon, F., and Bibette, J. (1998) Some stability criteria for double emulsions. *Langmuir* **14**, 2702–2706.
23. Florence, A. T., and Whitehill, D. (1981) Some feature of breakdown in water-in-oil-in-water multiple emulsions. *J. Colloid Interface Sci.* **79**, 243–256.
24. Shum, H. C., Lee, D., Yoon, I., Kodger, T., and Weitz, D. A. (2008) Double emulsion templated monodisperse phospholipid vesicles. *Langmuir* **24**, 7651–7653.
25. Wang, P., Sergeeva, M. V., Lim, L., and Dordick, J. S. (1997) Poly(methyl methacrylate) hollow particles by water-in-oil-in-water

emulsion polymerization. *Nat. Biotechnol.* **15**, 789–793.

26. Omi, S., Katami, K. I., Taguchi, T., Kaneko, K., and Iso, M. (1995) Synthesis of uniform PMMA microspheres employing modified SPG (Shirasu Porous Glass) emulsification technique. *Macromol. Symp.* **92**, 309–320.
27. Omi, S., Katami, K. I., Taguchi, T., Kazuyoshi, K., and Iso, M. (1995) Synthesis and applications of porous SPG (Shirasu Porous Glass) microspheres. *J. Appl. Polym. Sci.* **57**, 1013–1024.
28. Omi, S. (1996) Preparation of monodisperse microspheres using the Shirasu porous glass emulsification technique. *Colloid Surf. A-Physicochem. Eng. Asp.* **109**, 97–107.
29. Zhou, W.-Q., Gu, T.-Y., Su, Z.-G., and Ma, G.-H. (2007) Synthesis of macroporous poly(glycidyl methacrylate) microspheres by surfactant reverse micelles swelling method. *Eur. Polym. J.* **43**, 4493–4502.
30. Wang, R.-W., Zhang, Y., Ma, G.-H., and Su, Z.-G. (2006) Modification of poly(glycidyl methacrylate-divinylbenzene) porous microspheres with polyethylene glycol and their adsorption property of protein. *Colloid Surf. B-Biointerfaces* **51**, 93–99.

Chapter 14

Engineering the Logical Properties of a Genetic AND Gate

Daniel J. Sayut, Yan Niu, and Lianhong Sun

Abstract

Synthetic biology promises to enhance our ability to control biological systems by creating a systematic approach for the construction of genetic circuits that reliably program cellular function. As part of this approach, efficient methods are needed for the tuning of genetic circuits so as to allow for optimization of a design despite varying cellular contexts and incomplete understanding of in vivo biological interactions. Here we outline an optimization method that we have used to improve the logical responses of a genetic AND logic gate derived from components of the LuxI–LuxR bacterial quorum-sensing system. Basing our approach on the idea of evolutionary design, we improved the properties of our genetic AND logic gate by using directed evolution and a two-step screening process to alter the activities of the LuxR transcriptional activator. Using this method, we were able to rapidly enhance the AND gate's logical responses and have increased the specificities of these responses by ~1.5-fold.

Key words: Synthetic biology, genetic circuits, LuxR, directed evolution.

1. Introduction

The ability to rapidly construct and optimize genetic circuits lies at the heart of the developing field of synthetic biology (1). While mathematical models are useful in predicting the theoretical behavior of a proposed circuit design (2), the successful implementation of a design is complicated by an imprecise knowledge of in vivo biological properties and by the presence of cellular noise. These factors can require that a given circuit design go through many rounds of optimization and redesign before a functional circuit is achieved for a given recombinant host.

A particularly attractive method for the optimization of a genetic circuit is the directed evolution of the circuit's components (3). Using directed evolution, specific component

P. Wang (ed.), *Nanoscale Biocatalysis*, Methods in Molecular Biology 743,
DOI 10.1007/978-1-61779-132-1_14, © Springer Science+Business Media, LLC 2011

properties that are critical for circuit function can be evolved while retaining the general properties and design of the circuit (4). This allows for the components of a circuit to be chosen based on their function, such as a response to a specific environmental signal, instead of their inherent properties. Alternatively, proteins with desired properties (DNA binding, stability, etc.) can be evolved to respond to novel signals (5). Directed evolution is also an efficient way to generate diversity in a protein's properties, allowing for detailed characterizations of how a circuit's responses are affected by changing component properties (6).

To demonstrate how directed evolution can be used to improve the properties of a genetic circuit, we describe the construction and optimization of a genetic AND logic gate based on a hybrid quorum-sensing promoter. Our interest in this system is due to the possibility of using logic gates based on quorum-sensing components as motifs for the integration of intercellular signals in complex genetic circuits. In addition, the use of quorum-sensing components dictates that the optimization of the AND gate be done while largely retaining the original circuit design. Therefore, this system provides an excellent example of how directed evolution can be used for the optimization of a genetic circuit.

2. Materials

2.1. Cell Lines and Plasmids

1. TOP10F′ *Escherichia coli* (Invitrogen, Carlsbad, CA): F′{lacIqTn10(TetR)} mcrA Δ(mrr-hsdRMS-mcrBC) Φ80lacZΔM15 ΔlacX74 recA1 araD139 Δ(ara-leu)7697 galU galK rpsL endA1 nupG.
2. Plasmid pLac-LuxR (p15a ori, Kan^r): expresses *luxR* from the *lac/ara-1* inducible promoter ($P_{lac/ara\text{-}1}$) (*see* **Note 1**).
3. Plasmid pLuxI-GFPuv (colE1 ori, Cam^r): expresses *gfpuv* from the *luxI* quorum-sensing promoter (P_{luxI}).
4. Plasmid pBla-LuxR (p15a ori, Kan^r): expresses *luxR* from the constitutive beta-lactamase promoter (P_{bla}) (*see* **Section 3.2**).
5. Plasmid pLuxI-GFPmut2 (colE1 ori, Cam^r): expresses *gfpmut2* from P_{luxI}.

2.2. Enzymes and Reagents

1. Restriction enzymes and T4 DNA ligase with appropriate reaction buffers (New England Biolabs (NEB), Beverly, MA). Stored at –20°C.

2. *Pfu* (Stratagene, La Jolla, CA) and *Taq* (Thermo Scientific, Waltham, MA) polymerases with standard buffers (come with enzymes). Stored at −20°C.
3. PCR nucleotide mix: 10 mM each of dATP, dCTP, dGTP, and dTTP in water (Promega, Madison, WI). Stored at −20°C.
4. Mutagenic dNTP mixture: 2 mM dGTP, 2 mM dATP, 10 mM dCTP, and 10 mM dTTP made from individual solutions of dNTPs (Promega, Madison, WI). Stored at −20°C.

2.3. Media and Solutions

1. Lysogeny broth (LB): 1% w/v Bacto tryptone, 0.5% w/v Bacto yeast extract, 171 mM NaCl, pH 7.0. Sterilized by autoclaving.
2. SOC broth: 2% w/v Bacto tryptone, 0.5% w/v Bacto yeast extract, 10 mM NaCl, 2.5 mM KCl, 10 mM $MgCl_2$, 20 mM glucose, pH 7.0. Glucose and $MgCl_2$ should be sterilized separately and added after media is autoclaved.
3. LB agar: LB + 1.5% w/v agar. Sterilized by autoclaving.
4. Isopropyl-β-D-thiogalactopyranoside (IPTG) (Fisher Scientific, Fair Lawn, NJ): 420 mM in water, filter-sterilized.
5. 3-Oxohexanoyl-homoserine lactone (OHHL) (Sigma-Aldrich, St. Louis, MO): 1 mM in ethyl acetate acidified with glacial acetic acid (0.01% v/v) (*see* **Note 2**).
6. Antibiotic stocks: Chloramphenicol (Cam), 170 μg/mL in ethanol; kanamycin (Kan), 50 μg/mL in water; filter-sterilized.
7. Phosphate-buffered saline (PBS): 137 mM NaCl, 2.7 mM KCl, 10 mM sodium phosphate dibasic, 2 mM potassium phosphate monobasic, pH 7.4. Sterilized by autoclaving.
8. $MgCl_2$ solution: 70 mM in water, filter-sterilized.
9. $MnCl_2$ solution: 5 mM in water, filter-sterilized.
10. Agarose gels: 0.8–1.5% genetic analysis-grade agarose (*see* **Note 3**) in 1x TAE (40 mM Tris–acetate, 1 mM ethylene diamine tetraacetic acid [EDTA], 0.5 μg/mL ethidium bromide).

2.4. Kits and Equipment

1. QIAquick PCR purification kit (Qiagen, Chatsworth, CA).
2. Gel DNA Recovery Kit (Zymo, Orange, CA).
3. 96-well plate reader for fluorescence measurements (SPECTRAmax Gemini XS, Molecular Devices, Sunnyvale, CA).
4. 96-well plate reader for absorbance measurements (μQuant Universal Microplate Spectrophotometer, Bio-Tek, Winooski, VT).

5. 96-well clear-bottom plates (Corning, Corning, NY).
6. Thermocycler (MJ Research, Watertown, MA).
7. Electroporation equipment (Gene Pulser II, Bio-Rad, Hercules, CA).
8. Gel electrophoresis system (Thermo Scientific OWL Separation Systems, Rochester, NY).

3. Methods

3.1. Construction of Hybrid luxI-lacO Promoter ($P_{luxI\text{-}lacO}$)

P_{luxI} is the native promoter of the LuxI–LuxR quorum-sensing system and is activated by the LuxR transcriptional activator in the presence of OHHL (7, 8). AND logic is established in P_{luxI} by adding a *lac* operator site (*lacO*) downstream of the –10 site, creating $P_{luxI\text{-}lacO}$ (*see* **Fig. 14.1a**). In the absence of IPTG, which inactivates the LacI repressor, binding of *lacO* by LacI results in repression of $P_{luxI\text{-}lacO}$ (9). Therefore, expression from $P_{luxI\text{-}lacO}$ requires induction with both OHHL and IPTG. To create $P_{luxI\text{-}lacO}$, P_{luxI} is amplified from pLuxI-GFPmut2 using a reverse PCR primer that incorporates the *lacO* site. After amplification, the product is ligated into digested pLuxI-GFPmut2, replacing P_{luxI}.

1. Amplify P_{luxI} using PCR:

   Sample: 1 μL of DNA template (pLuxI-GFPmut2, 50 ng/μL), 10 μL 10x PFU buffer, 2 μL PCR nucleotide mix, 2.5 μL forward primer (100 ng/μL), 2.5 μL reverse primer (100 ng/μL), 2 μL *Pfu* DNA polymerase (2.5 U/μL), 80 μL sterilized water.

   Thermocycler Program: 35 cycles of 94°C for 45 s (denaturing), 53°C for 30 s (annealing), and 72°C for 30 s

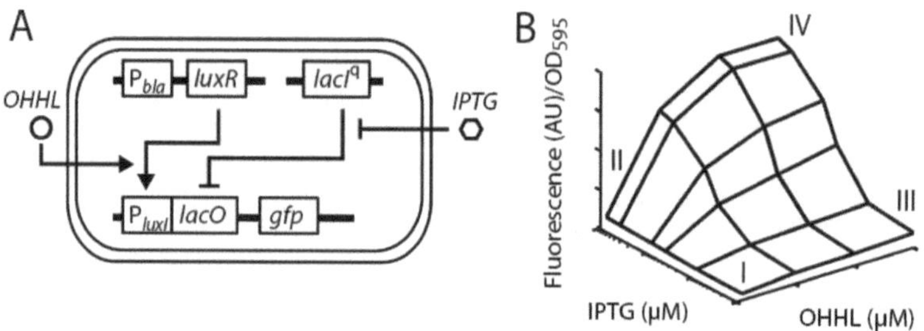

Fig. 14.1. Design and measurement of genetic AND logic gate. (**a**) Circuit design for establishment of AND logic using $P_{luxI\text{-}lacO}$. (**b**) Expected response for the AND logic gate for induction with IPTG and OHHL. Four expression states are identified: repressed basal (I, –IPTG/–OHHL); unrepressed basal (II, +IPTG/–OHHL); repressed active (III, –IPTG/+OHHL); and active (IV, +IPTG/+OHHL).

(extension) with an initial denaturing step (1 min) and a final extension step (10 min).

Forward Primer (*Aat*II): ACGGACGTCAGTCCTTTGATTCTAATAAATTGG

Reverse Primer (*Hin*dIII, *lac operator site*): ATTAAGCTT*GGAATTGTTATCCGCTCACAATTCC*ATTCGACTATAACAAAC

2. Purify amplified product with QIAquick PCR purification kit.
3. Digest amplified product (insert) and purified pLuxI-GFPmut2 (vector) with *Aat*II and *Hin*dIII restriction enzymes as per the manufacturer's instructions (*see* **Note 4**).
4. Purify digested insert and vector by separating on a 1.5% agarose gel and extracting desired bands (*see* **Note 5**). Applied voltage should be between 1 and 5 V/cm (distance between positive and negative electrodes). Correct bands are identified by comparing to a standard DNA ladder and are excised from the gel using a razor blade. After excision, the DNA is recovered using the Zymo Gel DNA Recovery Kit.
5. Ligate vector and insert using the following conditions: 100 ng digested vector, 3:1 ratio of insert/vector (*see* **Note 6**), 1 μL 10x DNA ligase buffer, 1 μL T4 DNA ligase, balance to 10 μL with water. Incubate in thermocycler at 16°C for 4–16 h.

3.2. Construction of pBla-LuxR

Plasmid pBla-LuxR is a derivative of pLac-LuxR that results in constitutive expression of LuxR from P_{bla}. Plasmid pBla-LuxR is constructed in an analogous manner to pLuxI(lacO)-GFPmut2 (*see* **Section 3.1**), with P_{bla} being amplified from a plasmid carrying the beta-lactamase gene (pUC19) and then ligated upstream of the LuxR gene in pLac-LuxR, replacing $P_{lac/ara\text{-}1}$. The primers used for amplification of P_{bla} are

Forward Primer (*Xho*I): TATCTCGAGCAGGTGGCACTTTTCG

Reverse Primer (*Eco*RI): GCGGAATTCTTTCAATATTATTGAAGC

Using these primers, the annealing temperature for the PCR reaction is 59°C and the extension time is 30 s.

3.3. Creation of AND Logic Circuit and Measurement of Logical Responses

The complete genetic circuit for the AND gate is achieved by co-transforming pLuxI(lacO)-GFPmut2 and pBla-LuxR into TOP10F′ *E. coli*. The circuit's logical responses are measured in 96-well plates using microplate readers.

1. Transform TOP10F′ *E. coli* with pLuxI(lacO)-GFPmut2 and pBla-LuxR (*see* **Note 7**) according to the manufacturer's instructions. After transformation, resuspend

transformed bacteria in 1 mL SOC and grow on a rotating incubator at 37°C and 225 rpm for 1 h prior to plating on LB plates containing Kan and Cam.

2. After overnight incubation of LB agar plates, use single colonies to inoculate culture tubes containing 2 mL of LB media (Kan^+,Cam^+) and grow overnight in a rotating incubator at 37°C and 225 rpm (*see* **Note 8**).
3. Inoculate at least 32 culture tubes containing 2 mL of fresh LB media (Kan^+, Cam^+) with 10 μL aliquots of the overnight cultures (16 tubes for each overnight culture, *see* **Note 9**).
4. Grow inoculated cultures for 3 h at 37°C and 225 rpm.
5. Induce tubes with varying concentrations of IPTG and OHHL. Inducer concentrations should be between 0 and 5 μM for OHHL and 0 and 100 μM for IPTG.
6. Continue incubation of cultures at 37°C and 225 rpm for 4 h.
7. Resuspend 1 mL of each sample in PBS using a microcentrifuge. Cell pellets are formed by centrifuging cultures at 10,000 rpm for 3 min. Pellets are resuspended gently in PBS by inverting microcentrifuge tubes.
8. Transfer 200 μL of each resuspended sample into wells of a 96-well plate.
9. Measure fluorescence (Ex. 481/Em. 507) and OD (595 nm) with microplate readers.
10. The logical properties of the AND gate are measured by comparing its expression levels at four different inductions as shown in **Fig. 14.1b**.

3.4. Optimization of AND Logic Using Directed Evolution of the LuxR Transcriptional Activator

Using a simple mathematical model it can be shown that the logical responses of the AND gate can be improved by altering the properties of LuxR for increased rates of transcriptional activation and, to a lesser degree, by increasing the amount of LuxR bound to the promoter at a given induction level (10). To enhance these properties, an error-prone PCR library of LuxR mutants is generated and then screened to identify mutants that increase the production of GFP in a minimal transcriptional circuit. To decrease the number of mutants that must be quantitatively screened, a qualitative plate-based fluorescence screen is used to identify putative positive mutants prior to quantitative screening. Once putative mutants are identified, their activities are then quantified to identify those that have the desired property of increased transcriptional activation in the limit of saturated OHHL induction. An overview of the screening process is presented in **Fig. 14.2**.

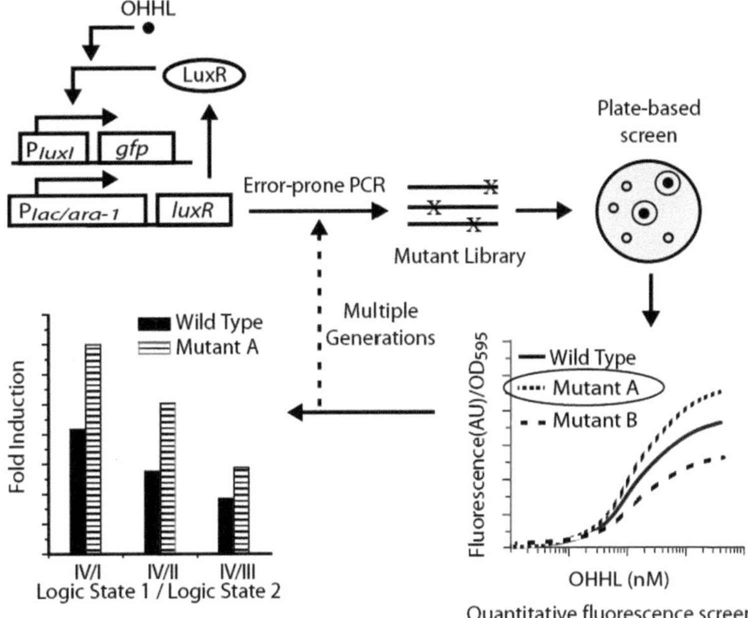

Fig. 14.2. Generation and screening of LuxR mutant proteins for improvement of genetic AND logic gate. *See* **Section 3.4** for details.

1. Create library of LuxR mutant proteins using error-prone PCR (11):

   Sample: 0.5 μL of DNA template (pLac-LuxR, 50 ng/μL), 10 μL 10x Taq buffer, 10 μL mutagenic dNTPs, 2.5 μL forward primer (100 ng/μL), 2.5 μL reverse primer (100 ng/μL), 4 μL $MnCl_2$ (5 mM) (*see* **Note 10**), 10 μL $MgCl_2$ (70 mM), 1 μL *Taq* DNA polymerase (5 U/μL), 59.5 μL sterilized water.

   Thermocycler Program: 30 cycles of 94°C for 45 s, 55°C for 30 s, and 72°C for 1.5 min with an initial denaturing step of 1 min.

   Forward Primer (*Kpn*I): GAAAGGTACCCATGAAAAACA TAAAT

   Reverse Primer (*Bam*HI): CGGGGATCCCGTACTTAATT
2. After amplification the mutant library and pLac-LuxR are digested with *Kpn*I and *Bam*HI, then purified using gel extraction (analogous to Steps 2–4 in **Section 3.1**).
3. Ligate mutant library into digested pLac-LuxR as described in Step 5 of **Section 3.1**.
4. Co-transform TOP10F′ *E. coli* with pLac-LuxR containing the mutant library and pLuxI-GFPuv (*see* **Note 11**). Cells are plated onto LB plates (Kan^+, Cam^+) treated with OHHL (200 nM) and IPTG (1 mM).

5. Incubate plates at 37°C for 18 h. After incubation, plates are kept at room temperature and examined for fluorescence development every hour. Colonies showing earlier fluorescence than wild-type controls are chosen for quantitative measurement as putative positive mutants.
6. Quantitative measurements of the putative LuxR mutants are done by following the same general method as outlined for the quantitative measurements of the genetic AND logic gate, but using TOP10F′ *E. coli* co-transformed with pBla-LuxR containing the mutant LuxR proteins and pLuxI-GFPmut2 instead of the AND gate plasmids (*see* **Section 3.3** and **Notes 12** and **13**).
7. LuxR mutants to be used for improvement of the AND gate's logical properties are distinguished by having increased expression levels at saturating concentrations of OHHL.
8. After identification of LuxR mutants (*see* **Note 14**), improved AND gates are constructed by co-transforming TOP10F′ *E. coli* with pBla-LuxR containing the mutant proteins and pLuxI(lacO)-GFPmut2.
9. If mutants identified during the first round of screening do not result in the desired properties, additional rounds (generations) of mutagenesis and screening may be done using the first generation mutants as templates.

4. Notes

1. Alternative plasmids may be used for all of the plasmids listed as long as the core components and regulatory functions are retained.
2. Prior to use, evaporate the ethyl acetate under a stream of air and dissolve the OHHL in sterile water.
3. The percentage of agarose depends on the size of the DNA fragments being separated. A 0.8% gel will resolve DNA in the range of 500 bp to 15 kb, while a 1.5% gel will resolve DNA in the range of 80 bp to 4 kb.
4. To overcome variations in the type and quality of DNA substrates, it is often necessary to use excess units of enzyme (between two- and tenfold) to achieve complete digestion.
5. While gel purification must be used to remove the digested fragment from the vector preparation, purification of the amplified insert can be done using the QIAquick kit due to the small size of the extraneous digested fragments.

6. The concentrations of the insert and vector can be estimated using gel electrophoresis and a DNA ladder with known molecular masses or by measuring absorbance at 260 nm using a spectrophotometer. The ratio of insert-to-vector can be calculated using the following formula: (size of insert/size of vector) $\times$ (concentration of vector) $\times$ 3 = mass of insert per volume of vector needed for a 3:1 ratio.
7. TOP10F′ is used because of the presence of the *lacI*^{q} gene on the F′ episome. Other strains may be used as long as there is a constitutively expressed copy of *lacI* present in the cell to repress expression from $P_{luxI\text{-}lacO}$.
8. At least two replicate overnight cultures should be used for quantification of the AND gate's responses to ensure reproducibility.
9. The number of culture tubes will depend on the desired number of measurements. Sixteen culture tubes will allow for four different concentrations of each of the inducers to be tested.
10. The mutation rate can be adjusted by changing the $MnCl_2$ concentration between 0.1 and 0.5 mM. For LuxR, a final $MnCl_2$ concentration of 0.2 mM resulted in an average of two amino acid substitutions per gene.
11. Plasmids pLuxI-GFPuv and pLac-LuxR are used in the screening of the LuxR mutants because the expression level of LuxR is increased for expression from $P_{lac/ara\text{-}1}$ in comparison to P_{bla}, and fluorescence development of GFPuv is delayed compared to GFPmut2. Combined, these properties simplify the screening process. It would also be possible to screen the LuxR mutants using pLuxI-GFPmut2 and pBla-LuxR.
12. The LuxR mutants can be transferred to pBla-LuxR by subcloning using the *Kpn*I and *Bam*HI restriction sites.
13. As P_{luxI} is used for measurement of the mutant responses, no IPTG needs to be added during induction.
14. LuxR mutants can be sequenced to identify the amino acid substitutions that result in the mutant's altered activities.

Acknowledgments

This work was supported by Grant # CBET-0747728 from the National Science Foundation and Grant # IRG 93-033 from the American Cancer Society.

References

1. Andrianantoandro, E., Basu, S., Karig, D. K., and Weiss, R. (2006) Synthetic biology: New engineering rules for an emerging discipline. *Mol. Syst. Biol.* **2**, 1–14.
2. Marguet, P., Balagadde, F., Tan, C., and You, L. (2007) Biology by design: Reduction and synthesis of cellular components and behaviour. *J. R. Soc. Interface* **4**, 607–623.
3. Haseltine, E. L., and Arnold, F. H. (2007) Synthetic gene circuits: Design with directed evolution. *Annu. Rev. Biophys. Biomol. Struct.* **36**, 1–19.
4. Yokobayashi, Y., Weiss, R., and Arnold, F. H. (2002) Directed evolution of a genetic circuit. *Proc. Natl. Acad. Sci. USA* **99**, 16587–16591.
5. Chockalingam, K., Chen, Z. L., Katzenellenbogen, J. A., and Zhao, H. M. (2005) Directed evolution of specific receptor-ligand pairs for use in the creation of gene switches. *Proc. Natl. Acad. Sci. USA* **102**, 5691–5696.
6. Alper, H., Fischer, C., Nevoigt, E., and Stephanopoulos, G. (2005) Tuning genetic control through promoter engineering. *Proc. Natl. Acad. Sci. USA* **102**, 12678–12683.
7. Whitehead, N. A., Barnard, A. M. L., Slater, H., Simpson, N. J. L., and Salmond, G. P. C. (2001) Quorum-sensing in gram-negative bacteria. *FEMS Microbiol. Rev.* **25**, 365–404.
8. Urbanowski, A. L., Lostroh, C. P., and Greenberg, E. P. (2004) Reversible acyl-homoserine lactone binding to purified Vibrio fischeri LuxR protein. *J. Bacteriol.* **186**, 631–637.
9. Gilbert, W., and Muller-Hill, B. (1970) The lactose repressor. In *The Lactose Operon* (Beckwith, J. and Zipser, D., Eds.), Cold Spring Harbor Lab Press, Plainview, NY, pp. 93–109.
10. Sayut, D. J., Niu, Y., and Sun, L. (2009) Construction and enhancement of a minimal genetic AND logic gate. *Appl. Environ. Microbiol.* **75**, 637–642.
11. Cadwell, R. C., and Joyce, G. F. (1994) Mutagenic PCR. *PCR Methods Appl.* **3**, S136–S140.

Chapter 15

Strain Engineering Strategies for Improving Whole-Cell Biocatalysis: Engineering *Escherichia coli* to Overproduce Xylitol as an Example

Jonathan W. Chin and Patrick C. Cirino

Abstract

This chapter provides an overview of key tools and methodologies available to practitioners of biocatalysis interested in using microorganisms to carry out biotransformations and describes specific examples of applying genetic modification strategies for strain design. We focus on the use of the polymerase chain reaction (PCR) for gene amplification, plasmid DNA for recombinant gene cloning and expression, and homologous recombination and phage transduction for modifying chromosomal DNA. Specifically we use *Escherichia coli* as the host organism, and the overproduction of xylitol by reduction of xylose represents the biotransformation of interest.

Key words: Polymerase chain reaction (PCR), P1 phage transduction, gene cloning, plasmid construction, gene knockout, xylitol, biocatalysis, *Escherichia coli*.

1. Introduction

Research in molecular biology has resulted in the development of a myriad of technologies for understanding and modifying cellular functions. Many fundamental genetic techniques have been developed into common tools and methodologies in biotechnology, enabling the engineering and design of cellular components and processes via genetic manipulation. Most of the common genetic engineering tools were developed in well-understood "workhorse" strains in molecular biology, e.g., *Escherichia coli* and *Saccharomyces cerevisiae*. While the genetic systems developed for these organisms are not universal to all microorganisms and often must be re-developed to work efficiently in other

P. Wang (ed.), *Nanoscale Biocatalysis*, Methods in Molecular Biology 743,
DOI 10.1007/978-1-61779-132-1_15, © Springer Science+Business Media, LLC 2011

organisms of interest, the concepts underlying the genetic modification strategies are generally universal. This chapter demonstrates the application of those concepts through specific examples and provides an overview of the key tools and methodologies available to practitioners of biocatalysis interested in using microorganisms to carry out biotransformations. We focus on the use of the polymerase chain reaction (PCR) for gene amplification, plasmid DNA for recombinant gene cloning and expression, and homologous recombination and phage transduction for modifying chromosomal DNA. Specifically we use *E. coli* as the host organism, and the overproduction of xylitol by reduction of xylose represents the biotransformation of interest.

Currently, xylitol is industrially produced by the catalytic hydrogenation of xylose at temperatures and pressures ranging from 80 to 140°C and ~50 atm, respectively, with a conversion of 50–60% (1). By using microorganisms as biocatalysts, xylitol can be produced in a potentially safer, more environmentally friendly process and achieve a higher product specificity. Some examples of xylitol bioproduction have been discussed previously (1, 2). Xylitol bioproduction, like many other valuable biotransformations, requires the use of a reduced cofactor (NAD(P)H) and therefore a means of regenerating these cofactors. While these reactions can be performed in vitro using purified enzymes, it is often easier and less expensive to regenerate reduced cofactors by harnessing sugar metabolism in whole cells for in vivo bioproduction (3, 4).

Although several yeast strains naturally produce xylitol (1, 2), our group has reported using engineered *E. coli* to produce xylitol from a glucose and xylose mixture (5–8). We use *E. coli* because it is a fast-growing, robust organism that is genetically easy to manipulate through gene insertion and deletion strategies and gene overexpression systems. However, *E. coli* does not naturally produce xylitol, nor does it readily co-utilize xylose and glucose. Therefore a number of genetic modifications were implemented to alter native metabolism, so it is better suited for xylitol production.

In this chapter, we describe the materials and methods used to construct an *E. coli* strain that efficiently produces xylitol from a glucose and xylose mixture when a NADPH-dependent xylose reductase from *Candida boidinii* (denoted *CbXR*) is heterologously expressed from a plasmid (7). We start by constructing a plasmid carrying the *CbXR* gene. The gene is amplified from *C. boidinii* genomic DNA using high-fidelity DNA polymerase chain reaction (**Section 3.1**) and then cloned under control of a p_{tac} promoter (inducible with isopropyl β-D-1-thiogalactopyranoside, IPTG) in a medium copy plasmid (**Section 3.1**) creating plasmid pCBXR. We then construct an *E. coli* strain in which the *xylB* gene encoding the xylulokinase enzyme (XylB) is disrupted, effectively disabling xylose metabolism (**Section 3.2**). In addition to ensuring that xylose is used solely for xylitol production and not cell growth, the *xylB* deletion additionally prevents toxic synthesis of xylitol phosphate by XylB (9) and increases NADPH availability (from glucose metabolism) for xylose reduction, relative to co-metabolism of both glucose and xylose. Lastly, in order to allow for xylose transport in the presence of glucose, we replace the native catabolite-sensitive regulatory protein CRP ("cyclic AMP receptor protein") (10) with a CRP mutant ("CRP*") that is able to activate catabolic operons (such as the *xyl* genes for xylose) in the presence of glucose (11). To achieve chromosomal replacement of the wild-type *crp* gene with *crp**, P1 phage transduction is used (**Section 3.5**). The resulting double mutant strain, when transformed with the *CbXR*-containing plasmid (pCBXR), is an effective xylitol-producing microorganism when growing on glucose in the presence of xylose (**Section 3.6**) (7).

2. Materials

2.1. Strains and Plasmids

1. Strain W3110 (ATCC 27325).
2. Strain DH5α (ATCC 53868) (*see* **Note 1**).
3. Strain ET23 (11). *crp**, streptomycin resistant (*rpsL* 150) and tetracycline resistant (*Tn10*).
4. P1 *vir* phage (common lab phage) (*see* **Note 2**).
5. Plasmid pLOI3421 (12). Carries apramycin resistance gene (*aac*) flanked by FRT recognition sequences.
6. Plasmid pCP20 (13). Used to express the Flp recombinase; ampicillin resistance (*bla*).
7. Plasmid pKD46 (14). Used to express the Red recombinase system; ampicillin resistance (*bla*).
8. Genomic DNA from yeast strain *C. boidinii* NRRL Y-17213 (15). This can be obtained from our lab by request or prepared as described by Sambrook and Russell (16).
9. Plasmid pLOI3809 (7). This is derived from plasmid pFLAG-CTC (Sigma-Aldrich, St. Louis, MO) but with the ampicillin resistance gene (*bla*) replaced by a kanamycin resistance gene (*kan*) (*see* **Note 3**).

2.2. Polymerase Chain Reaction (PCR) Reagents

1. Appropriate PCR primers diluted to 100 μM. Store at −20°C (*see* **Note 4**).
2. Deoxynucleotide solution set (dNTPs diluted to a 10 mM stock mixture). Store at −20°C (New England BioLabs (NEB), Beverly, MA).

3. DNA Polymerase (*Taq* (NEB, Beverly, MA) for standard analytical PCR and Phusion (NEB, Beverly, MA) for high-fidelity PCR). Store at −20°C (*see* **Note 5**).
4. 5x or 10x polymerase reaction buffer (supplied with DNA polymerase). Store at −20°C.

2.3. Culturing Reagents

1. Antibiotics at 1,000x concentrated stock solution: Apr (apramycin, 50 mg/mL), Amp (ampicillin, 100 mg/mL), Kan (kanamycin, 50 mg/mL), Tet (tetracycline, 50 mg/mL), Str (streptomycin, 50 mg/mL). Store at −20°C until use. *Caution*: tetracycline is light sensitive!
2. LB medium (10 g/L tryptone, 5 g/L yeast extract, and 5 g/L NaCl in double distilled water (ddH_2O)): antibiotics can be added as needed to get a 1x final concentration.
3. LB plates (10 g/L tryptone, 5 g/L yeast extract, 5 g/L NaCl, and 15 g/L agar in ddH_2O): antibiotics can be added as needed to get a 1x final concentration.
4. SOB medium (20 g tryptone and 5 g yeast extract in 0.975 L ddH_2O). After autoclaving, add 5 M NaCl to 10 mM final concentration, 1 M KCl to 2.5 mM final, 1 M $MgSO_4$ to 10 mM final, and 1 M $MgCl_2$ to 10 mM final (all reagent stocks should be previously autoclave-sterilized).
5. SOC medium: to prepare, add 0.725 mL of pre-sterilized 50% glucose (w/v) to 100 mL SOB, resulting in 20 mM glucose final concentration.

2.4. Plasmid Construction Reagents

1. 0.8% agarose gel (prepared as 0.8% w/v agarose in 1x TAE buffer). TAE: 40 mM Tris–acetate, 1 mM EDTA (*see* **Note 6**).
2. 1x TAE solution containing 0.5 μg/mL ethidium bromide (EtBr) (*see* **Note 7**).
3. PCR clean-up kit (e.g., Qiaquick PCR Purification Kit, Qiagen, Valencia, CA, or DNA Clean & Concentrator-5 Kit, Zymo Research, Orange, CA).
4. Gel extraction clean-up kit (e.g., Qiaquick Gel Extraction Kit, Qiagen, Valencia, CA, or Zymoclean Gel DNA Recovery Kit, Zymo Research, Orange, CA).
5. Appropriate restriction endonucleases (*Nde*I, *Xho*I, *Dpn*I (NEB, Beverly, MA)). Store at −20°C.
6. 10x restriction endonuclease reaction buffer (supplied with each enzyme). Store at −20°C.
7. T4 DNA ligase (Invitrogen, Carlsbad, CA). Store at −20°C.
8. 5x T4 DNA ligase reaction buffer (supplied with ligase). Aliquot into smaller volumes (~10–20 μL) and store at

−20°C. Repeated freeze/thaws are detrimental to the stability of ATP included in the reaction buffer.

9. Nitrocellulose dialysis membrane (0.025 μm VSWP MF^{TM} membrane filters, Millipore, Billerica, MA). *Caution*: nitrocellulose membrane is flammable!
10. Plasmid miniprep kit (e.g., Qiaprep Spin Miniprep Kit, Qiagen, Valencia, CA, or Zyppy Plasmid Miniprep Kit, Zymo Research, Orange, CA).

2.5. Gene Knockout Reagents

1. 1 M L-arabinose dissolved in ddH_2O.

2.6. Phage Transduction Reagents

1. 0.1 M $CaCl_2$ dissolved in ddH_2O.
2. $CHCl_3$; *Caution*: use in a properly ventilated area!
3. 1.0 M $MgSO_4$ dissolved in ddH_2O.
4. 1.0 M Na-citrate dissolved in ddH_2O.

3. Methods

3.1. Gene Amplification and Cloning

This section provides an overview on the use of high-fidelity DNA polymerase to amplify a specific gene (*CbXR*) from a genomic prep. The gene is then cloned into an expression plasmid (pLOI3809). The primers used for gene amplification are designed to contain a selected pair of restriction sites (*Nde*I and *Xho*I) which can then be digested and subsequently ligated at a specific location in a vector (pLOI3809), which has also been digested with the same restriction enzymes. Refer to **Note 4** for more information regarding cloning primers.

1. From a genomic DNA prep of *C. boidinii* and using primers ***CbXR* for** and ***CbXR* rev**, prepare a high-fidelity polymerase chain reaction (PCR) mixture to amplify the xylose reductase gene (*CbXR*). For a 50 μL reaction, add 34 μL H_2O, 10 μL 5x Phusion buffer, 1 μL 10 mM dNTPs, 2.5 μL 10 μM primer mixture, ~10 ng template DNA (genomic prep of *C. boidinii*), and 0.5 μL Phusion enzyme (the enzyme should be added last and kept at −20°C until use) (*see* **Note 6**). The primer mixture contains the ***CbXR* for** and ***CbXR* rev** primers at a concentration of 10 μM.
2. The PCR reaction should then be performed using the following thermocycler protocol: Heat at 98°C for 30 s; 25 cycles of 98°C for 10 s, 69°C for 20 s, 72°C for 20 s; followed by a final extension of 72°C for 5 min; then hold at 4°C. This protocol is specific to the Phusion

enzyme amplifying ~1 kb fragment using primers ***CbXR* for** and ***CbXR* rev**. The resulting PCR product is the *CbXR* gene flanked by the restriction sites *Nde*I and *Xho*I, which were designed into the primers. The PCR product is next purified from the PCR mixture using gel electrophoresis and then digested using restriction enzyme endonucleases to create sticky ends for directional ligation into plasmid pLOI3809.

3. Prepare a 0.8% agarose gel (0.8% w/v agarose in 1x TAE) and place into a gel box containing 1x TAE with 0.5 μg/mL ethidium bromide (EtBr) (*see* **Note 7**).
4. Add 10 μL 6x loading buffer to the 50 μL PCR reaction mixture and load the buffer+PCR mixture into the gel's sample well.
5. Set the gel electrophoresis power pack to a constant voltage of 90 V and run the gel electrophoresis for ~45 min.
6. Excise the PCR product (~1 kb) from the gel while visualizing the DNA over a UV box, taking care to minimize UV exposure to the DNA and user. Proper protection must be worn to minimize direct exposure to UV.
7. Clean the PCR product from the gel fragment using a gel extraction clean-up kit, following the manufacturer's protocol.
8. Perform restriction digests of the cleaned PCR product DNA (this is the "insert" gene for the ligation) and plasmid pLOI3809 (vector) using the restriction endonucleases *Nde*I and *Xho*I. A single reaction can include both enzymes when using NEB buffer 4 and a 1x concentration of BSA (supplied by the manufacturer) (this is called a "double digest"). Typical reaction time is ~3 h at 37°C. The resulting products are the *CbXR* gene and pLOI3809 vector flanked on either side by sticky ends corresponding to the *Nde*I and *Xho*I restriction sites.
9. Clean the digested products using a PCR or DNA clean-up kit, following the manufacturer's protocol. This removes unwanted salts and proteins that may interfere with the ligation reaction.
10. Perform a ligation reaction to clone the digested *CbXR* gene into the digested pLOI3809 vector using T4 DNA ligase. Follow the manufacturer's protocol for buffer, vector, and insert concentrations (*see* **Note 8**). The resulting product is the *CbXR* gene ligated into plasmid vector pLOI3809: "pCBXR."
11. The ligation reaction mixture containing the desired cloned plasmid also contains a high salt concentration that is

detrimental to electroporation. The salt concentration can be diluted by performing a dialysis step. This is accomplished by placing a nitrocellulose filter "shiny side up" in a petri dish filled with ddH_2O. Place ~12 μL of the ligation product onto the filter by pipette for ~1 h (~100–200 ng DNA). Carefully remove the ligation product from the filter by pipette and transfer the product to an ice-cold, pre-sterilized microcentrifuge tube (*see* **Note 9**).

Next, the cloned plasmid must be transformed into *E. coli* (*see* **Note 10**). Before using the plasmid for its intended application, it is important to first verify that the gene has been correctly cloned. This involves isolating a single cloned colony carrying the transformed plasmid, growing up the cells to amplify the plasmid, and then purifying a quantity of the plasmid DNA sufficient for analysis. These steps are detailed below.

12. Prepare electrocompetent *E. coli* strain DH5α. This is a common lab strain used for carrying out standard cloning steps such as transforming and amplifying ligation products (*see* **Note 10**).
13. Add 70 μL of the electrocompetent cell suspension to the microcentrifuge tube containing the dialyzed ligation product. Mix and transfer the entire contents to an ice-cold sterile electroporation cuvette.
14. Electroporate according to standard protocols (12.5–15 kV/cm, for 4.5–5.5 ms, in a cuvette with a 0.1–0.2 cm gap length (16)) and quickly add ~0.8 mL SOC medium immediately following the voltage pulse.
15. Transfer the SOC + cell mixture to a culture tube and incubate at 37°C with 250 rpm shaking for 45–60 min. This step, called the outgrowth, allows the cells to grow and express the antibiotic resistance marker without selection pressure of an antibiotic. However, longer incubation times may result in plasmid loss due to a lack of selection pressure.
16. Spread the outgrowth medium, at various dilutions, onto multiple LB + Kan plates and incubate overnight at 37°C.
17. Isolate single colonies from the LB + Kan plates and prepare overnight cultures by inoculating 3 mL LB + Kan medium with the single colonies (one colony per culture). Incubate these cultures at 37°C with 250 rpm shaking.
18. Purify the plasmid DNA from the overnight cultures using a standard "miniprep" kit, following the manufacturer's protocol.
19. Verify that the *CbXR* gene was successfully cloned into the pLOI3809 vector. This can be accomplished by restriction

digest analysis using the enzyme *Bam*HI. The resulting digestion pattern should show two bands (sized 3.2 and 4.2 kb) when analyzed by gel electrophoresis. DNA sequencing of the plasmid can of course also be used to confirm proper cloning. The resulting plasmid containing the *CbXR* gene is pCBXR. Minipreps of confirmed pCBXR DNA from strain DH5α can now be used to transform the plasmid into other strains.

3.2. Deleting the xylB Gene from the W3110 Chromosome

This section provides an overview on the use of homologous recombination to disrupt a chromosomal gene. The protocol takes advantage of the phage lambda Red recombinase, which mediates recombination between homologous sequences, and the site-specific FLP recombinase, which acts on the 34-base pair FLP recognition target (FRT) sites. First, an FRT-flanked apramycin resistance gene (*aac*) is amplified from plasmid pLOI3421 using PCR. The PCR primers are designed such that approximately 30–50 base pairs at the 5′-end of each primer is homologous to either end of the gene to be disrupted (*xylB*), while the 3′-ends are homologous to plasmid pLOI3421. The resulting PCR product is the FRT-flanked *aac* gene flanked by regions of homology to the *xylB* gene. This linear DNA fragment is transformed into the strain intended to receive the *xylB* knockout. The Red recombinase system is expressed in the strain to ensure double recombination between the genomic *xylB* gene and the region of homology on the PCR product, which represents the desired gene disruption that is easily selected for by apramycin resistance.

1. From a plasmid miniprep of pLOI3421 and using primers **xylB for KO** and **xylB rev KO**, perform a standard PCR to amplify FRT-flanked *aac* region (~2.1 kb). For a 50 μL reaction, add 36.5 μL H_2O, 5 μL 10x polymerase buffer, 2 μL 10 mM dNTPs, 4 μL 10 μM primer mixture, 2 μL template DNA (~0.5 μg from a plasmid miniprep of pLOI3421), and 0.5 μL *Taq* polymerase enzyme (the enzyme should be added last and kept at −20°C until use) (*see* **Note 11**). The primer mixture contains the **xylB for KO** and **xylB rev KO** primers at a concentration of 10 μM.
2. Run the thermocycler with the following protocol. Heat at 95°C for 4 min; 30 cycles of 95°C for 45 s, 56°C for 45 s, 72°C for 132 s; followed by a final extension of 72°C for 5 min; then hold at 4°C. This protocol is specific for using a standard *Taq* polymerase amplifying a 2.1 kb fragment using the primer pair **xylB for KO** and **xylB rev KO**. The resulting PCR product is the FRT-flanked apramycin resistance gene flanked by 50 base pairs homologous to the *xylB* gene. However, the PCR reaction mixture still contains the template plasmid pLOI3421, which will confer

apramycin resistance to the recipient strain if transformed. In order to eliminate these false-positive transformants the PCR mixture must be treated with *Dpn*I endonuclease, which digests methylated DNA (plasmid) but not PCR product.

3. Digest the PCR mixture with the restriction endonuclease *Dpn*I. This can be performed by adding 2 μL of *Dpn*I directly to the 50 μL PCR mixture (*see* **Note 12**). Typical reaction time is ~3 h at 37°C. The resulting products are the uncleaved PCR product and digested plasmid (methylated DNA).
4. Perform gel electrophoresis and gel extraction as outlined earlier (Steps 3 and 7 of **Section 3.1**) to isolate, excise, and purify the ~2.1 kb PCR product from the PCR/*Dpn*I digest mixture.
5. Prepare electrocompetent *E. coli* strain W3110 to transform with plasmid pKD46 (Amp^R) (*see* **Note 10**). Transformation of plasmid pKD46 into W3110 allows the genes encoding the Red recombinase system to be heterologously expressed in W3110. Expression of the recombinase is induced by L-arabinose. Replication of pKD46 is temperature sensitive, so that strains are easily cured of the plasmid by growing 37–42°C.
6. In an ice-cold pre-sterilized microcentrifuge tube add plasmid pKD46 DNA (~1 ng) followed by 70 μL of the electrocompetent W3110 cell suspension. Mix and transfer the entire contents to an ice-cold sterile electroporation cuvette.
7. Electroporate according to standard protocols (12.5–15 kV/cm, for 4.5–5.5 ms, in a cuvette with a 0.1–0.2 cm gap length (16)) and quickly add ~0.8 mL SOC medium immediately following the voltage pulse.
8. Transfer the SOC + cell mixture to a culture tube and perform the outgrowth at 30°C with 250 rpm shaking for 45–60 min. *Caution*: plasmid pKD46 is temperature sensitive and cannot be replicated at 37°C.
9. Spread dilutions of the outgrowth cell mixture onto LB+Amp plates and incubate overnight at 30°C.
10. Prepare an overnight culture of W3110+pKD46 by inoculating 3 mL LB+Amp with a single colony from the LB+Amp plate and incubate at 30°C with 250 rpm shaking.
11. Inoculate 30 mL of SOB with ~3–5 drops of the overnight culture (W3110 + pKD46). Supplement the SOB medium with ampicillin (for plasmid selection) and 1 mM

L-arabinose (for expression of the Red recombinase system) and grow at 30°C with 250 rpm shaking to an OD_{600} ~ 0.6 (*see* **Note 13**).

12. Prepare electrocompetent W3110 + pKD46 cells to transform with the PCR product containing the FRT-flanked apramycin resistance gene (*see* **Note 10**).
13. Transform 70 μL of electrocompetent W3110 + pKD46 cells with ~200 ng of the purified PCR product from Step 4 of **Section 3.2** (*see* **Note 14**) and electroporate according to standard protocols (12.5–15 kV/cm, for 4.5–5.5 ms, in a cuvette with a 0.1–0.2 cm gap length (16)) and quickly add ~0.8 mL SOC medium immediately following the voltage pulse.
14. Transfer the SOC medium + cell mixture to a culture tube and perform the outgrowth at 37°C with 250 rpm shaking for 45–60 min.
15. Spread approximately half of the outgrowth medium, at various dilutions, onto multiple LB + Apr plates and incubate overnight at 37°C (*see* **Note 15**).
16. From the LB + Apr plates, isolate single colonies and verify the *xylB* gene deletion by performing an "analytical" PCR using *Taq* polymerase and the **xylB for check** and **xylB rev check** primers. The template DNA can be prepared directly from a colony by suspending a single colony into ~30 μL ddH_2O and heating the cell solution at 95°C for 5 min. For a 50 μL reaction, add 36.5 μL H_2O, 5 μL 10x polymerase buffer, 2 μL 10 mM dNTPs, 4 μL 10 μM primer mixture, 2 μL template DNA (from the heated cell solution), and 0.5 μL *Taq* enzyme (the enzyme should be added last and kept at −20°C until use). The primer mixture contains the **xylB for check** and **xylB rev check** primers at a concentration of 10 μM.
17. Run the thermocycler with the following protocol. Heat at 95°C for 4 min; 30 cycles of 95°C for 45 s, 56°C for 45 s, 72°C for 132 s; followed by a final extension of 72°C for 5 min; then hold at 4°C. The resulting PCR product amplifies the *xylB* region and will be either ~1.4 kb (if the double recombination event did not occur) or ~2.1 kb (if the desired recombination event occurred and the amplified region contains the FRT-*aac*-FRT sequence).
18. From the single colony that has been verified to contain the desired gene deletion, ensure that the plasmid has been cured of pKD46 by testing for growth on an LB + Amp at 30°C. If growth occurs, streak the desired Apr^R colony from the LB + Apr plate onto another LB + Apr plate and incubate overnight at 42°C. Verify plasmid loss by streaking

single colonies onto LB + Amp plates, incubating overnight at 30°C, and checking for growth. W3110 with Δ*xylB* and apramycin resistance = strain "WΔB *aac*." The deleted *xylB* gene is now replaced by the FRT-flanked *aac* gene. This disruption prevents xylose metabolism and can be verified by the strain's inability to grow with xylose as the only carbon source.

3.3. Remove FRT-Flanked Antibiotic Resistance (Apr^R)

This section provides an overview of the use of FLP recombinase to remove the FRT-flanked apramycin resistance gene (*aac*), which has replaced the deleted *xylB* fragment. The resulting recombination product is the apramycin-sensitive Δ*xylB* strain containing a FRT "scar" in the *xylB* region.

1. Prepare an overnight culture by inoculating 3 mL LB medium with a single colony of WΔB *aac* and incubate at 37°C with 250 rpm shaking.
2. Prepare electrocompetent *E. coli* strain WΔB *aac* cells to transform with plasmid pCP20 (*see* **Note 10**). pCP20 is a temperature-sensitive plasmid that carries FLP recombinase under the control of a temperature-inducible promoter (13).
3. Transform 70 μL of electrocompetent WΔB *aac* cells with ~1 ng of plasmid pCP20 and electroporate according to standard protocols (12.5–15 kV/cm, for 4.5–5.5 ms, in a cuvette with a 0.1–0.2 cm gap length (16)) and quickly add ~0.8 mL SOC medium immediately following the voltage pulse.
4. Transfer the SOC medium + cell mixture to a culture tube and perform the outgrowth at 30°C with 250 rpm shaking for 45–60 min. *Caution*: plasmid pCP20 is temperature sensitive and cannot be replicated at 37°C.
5. Spread dilutions of the outgrowth cell mixture onto LB + Amp plates and incubate overnight in a 30°C.
6. Isolate a single colony from the LB + Amp plate, streak onto an LB plate, and incubate at 42°C overnight. This serves the double purpose of inducing expression of the FLP recombinase and curing the strain of plasmid pCP20.
7. Isolate single colonies from the LB plate and verify the loss of both plasmid pCP20 and FRT-flanked apramycin resistance by streaking single colonies onto LB + Amp, LB + Apr, and LB plates. Colonies that have lost the ability to grow on Amp have been cured of plasmid pCP20, while colonies that cannot grow on Apr have lost the *aac* gene via FRT recombination. W3110 with Δ*xylB* without apramycin resistance = WΔB aac^- = WΔB.
8. Verify the apramycin resistance gene removal by performing a standard PCR using the **xylB for check** and **xylB rev**

check primers, as described earlier (Step 16 of **Section 3.2**). The resulting PCR product amplifies the *xylB* region and will be either ~2.1 kb (if the recombination event did not occur (strain WΔB *aac*+)) or ~0.4 kb (if the desired recombination event occurred, resulting in strain "WΔB"). As before, this strain should not be able to metabolize xylose.

3.4. Preparation of a P1 Phage Lysate

P1 phages being assembled in an infected *E. coli* host strain will occasionally package host chromosomal fragments of ~90 kb, rather than phage DNA. This section describes the preparation of a lysate of P1 phage carrying a library of packaged "donor" strain genomic DNA. In this case, the donor is strain ET23 (11), which carries the *crp** gene flanked on one end by a streptomycin resistance marker (*rpsL* 150) and on the other end by a tetracycline resistance marker (Tn10) (11). Refer to (16, 17) for more detailed descriptions of phage transduction applications and protocols.

1. Prepare an overnight culture by inoculating 3 mL LB medium with the host strain (ET23) and incubate at 37°C with 250 rpm shaking.
2. Dilute the overnight culture 1:100 in 15 mL LB (3 drops in 15 mL) in two separate 250 mL Erlenmeyer flasks. One flask will later be infected with phage; the second flask serves as a control to check if the phage infection was successful.
3. Grow the 15 mL cultures to an $OD_{600} \sim 0.2$ (~2 h) at 37°C.
4. Add 1.6 mL 0.1 M $CaCl_2$ to both cultures to achieve a final concentration of 10 mM.
5. Add 150 μL of the original stock of P1 *vir* phage lysate (should contain $\sim 1 \times 10^8$ phage particles per mL) to one of the cultures.
6. Grow both cultures for another 4–6 h. After addition of the phage, you may see small clumps of cell debris. The lysed culture should appear much clearer than the uninfected control. This indicates phage replication and cell lysis.
7. Add 0.5 mL $CHCl_3$ to the infected culture using a pre-sterilized glass pipette.
8. Transfer the infected culture to an ice-cold, sterile 15 mL conical/centrifuge tube and invert several times. If you are using a plastic centrifuge tube, you need to work quickly as chloroform will begin to dissolve the plastic.
9. Centrifuge the tube at 3,750 rpm for 10 min at 4°C to pellet the cellular debris.
10. Carefully transfer the supernatant, which contains the phage lysate, to a pre-sterilized 15 mL glass tube

containing 0.5 mL $CHCl_3$. This P1 phage lysate can be stored at 4°C until use. The lysate represents a phage-packaged library of ET23 chromosomal DNA including the mutant *crp* gene (*crp**) flanked by tetracycline and streptomycin resistance selection markers.

3.5. P1 Phage Transduction

Infecting a recipient strain with the phage lysate allows for introduction of the donor chromosomal DNA fragments into the recipient strain ("transduction"). Homologous recombination allows the recipient genomic DNA to be replaced by the phage-introduced donor DNA. Selection for the desired recombination event is accomplished via genetic markers neighboring the gene of interest (e.g., antibiotic resistance). In our example, strain WΔB is transduced with DNA from strain ET23, and recombinant transductants carrying the desired *crp** gene mutations are isolated by selecting for strains having streptomycin and tetracycline resistance. This section describes the phage transduction protocol.

1. Prepare an overnight culture of the recipient strain (WΔB) by inoculating a 3 mL culture with a single colony of WΔB and incubate at 37°C with 250 rpm shaking.
2. Aliquot 1 mL of the overnight culture into a sterile microcentrifuge tube and centrifuge for 5 min at 5,000 rpm and room temperature.
3. Concentrate the cells twofold by removing 500 μL of the supernatant and resuspending the cells in the leftover supernatant.
4. Add 25 μL 0.1 M $CaCl_2$ and 5 μL 1 M $MgSO_4$ to the cell suspension and mix by inversion.
5. Prepare five sterile microcentrifuge tubes containing the following:

Tube	WΔB cells (mL)	ET23 P1 lysate (μL)
1	0.1	
2	0.1	5
3	0.1	10
4	0.1	50
5	–	50

6. Incubate the tubes for 30 min at 37°C.
7. Add 0.1 mL 1 M Na-citrate to each tube and mix by inversion.
8. Add 1 mL LB to each tube, mix by inversion, and incubate for 1–1.5 h at 37°C.

9. Centrifuge the tubes for 5 min at 5,000 rpm and room temperature.
10. Remove most of the supernatant, leaving ~150 μL of the supernatant.
11. Resuspend the cell pellet with the remaining 150 μL supernatant and spread the cell suspension onto an LB+Str+Tet plate, spreading the mixture evenly across the plate. Incubate the plates overnight at 37°C.
12. Isolate single colonies from the plates and verify that the *crp* gene was replaced by *crp**. Verification can be accomplished either by (i) growing the strain in a medium supplemented with glucose and arabinose and then analyzing the culture broth by HPLC to detect simultaneous glucose and arabinose consumption or by (ii) performing a triphenyltetrazolium chloride (TTC) assay for expression of arabinose catabolism genes in the presence of glucose (*see* **Note 16**).

    The final strain contains both the *xylB* gene deletion and the *crp** mutation (strain "WΔB*"). This strain is able to simultaneously uptake glucose and xylose, due to the *crp** mutation, and prevents xylose from being metabolized, due to the *xylB* gene disruption. Both of these mutations are important for xylitol production.

3.6. Xylitol Production

This section describes the procedure of producing xylitol in a batch culture using strain WΔB* and plasmid pCBXR.

1. Prepare an overnight culture of WΔB* by inoculating a 3 mL LB culture with a single colony of WΔB* and incubate at 37°C with 250 rpm shaking.
2. Prepare electrocompetent *E. coli* strain WΔB* cells to transform with the plasmid pCBXR (*see* **Note 10**). Transformation of plasmid pCBXR into WΔB* allows for heterologous expression of *CbXR*.
3. Transform 70 μL of electrocompetent WΔB* cells with ~1 ng of plasmid pCBXR and electroporate according to standard protocols (12.5–15 kV/cm, for 4.5–5.5 ms, in a cuvette with a 0.1–0.2 cm gap length (16)) and quickly add ~0.8 mL SOC medium immediately following the voltage pulse.
4. Transfer the SOC medium + cell mixture to a culture tube and perform the outgrowth at 37°C with 250 rpm shaking for 45–60 min.
5. Spread dilutions of the outgrowth cell mixture onto LB + Kan plates at various dilutions and incubate overnight at 37°C.

6. From the LB + Kan plate, isolate a single colony (WΔB*+pCBXR) and streak onto another LB + Kan plate and incubate at 37°C overnight.
7. From the fresh LB + Kan plate, prepare an overnight culture by inoculating 3 mL LB + Kan with a single colony of WΔB*+pCBXR and incubate at 37°C with 250 rpm shaking.
8. Prepare LB medium supplemented with 300 mM xylose, 100 mM glucose, 50 mM MOPS (pH 7.4), and 50 μg/mL Kan; this medium will be referred to as LBXGM.
9. Use the overnight culture of WΔB*+pCBXR to inoculate 50 mL of LBXGM medium to an $OD_{600} = 0.1$. Add 100 μM IPTG to induce *CbXR* expression. Incubate the culture at 30°C with 250 rpm shaking. Sugar consumption and xylitol production can be monitored via HPLC. The culture should convert all xylose to xylitol, as reported (7).

4. Notes

1. DH5α is a standard molecular cloning strain that, due to several genetic alterations, shows increased transformation efficiency, increased plasmid yield and purity, and increased stability (18).
2. P1 *vir* phage is a common lab phage that enters the lytic cycle upon infection (19).
3. There are many expression vectors available with different features such as origin of replication, copy number, antibiotic/selection marker, promoter/inducible gene expression system, temperature sensitivity. Copy number determines the number of plasmids contained per cell and ranges between low (1–5 copies per cell), medium (15–50 copies per cell), and high (500–700 copies per cell) (16). Antibiotics that are commonly used for plasmid maintenance in *E. coli* (with their corresponding genes encoding resistance) include kanamycin (*kan*), ampicillin (*bla*), apramycin (*aac*), tetracycline (*Tn10*), and carbenicillin (*bla*) (16). Some plasmids are temperature sensitive and are not replicated above ~30°C (e.g., pKD46 and pCP20 described in this chapter). This makes it simple to "cure" a strain of these plasmids by incubating at elevated temperatures. Promoter systems vary by strength, stringency, and method of induction. Refer to (20, 21) for a detailed review of protein expression vectors.

The pFLAG-CTC expression vector (Sigma-Aldrich, St. Louis, MO) that was used to construct plasmid pLOI3809 described in this chapter is a medium copy plasmid containing a pBR322 origin of replication. The plasmid contains a ribosomal binding site upstream of the start codon/cloning site, and gene expression is under control of the *tac* promoter (induced with IPTG).

4. Refer to Sambrook and Russell for detailed primer design considerations (16). Primers used in this chapter are as follows:

   xylB for KO: 5′-CCG GTT ATC GGT AGC GAT ACC GGG CAT TTT TTT AAG GAA CGA TCG ATA TG GGC GAT TAA GTT GGG TAA CG-3′ ($T_{m,xylB}$ = 77°C, $T_{m,3421}$ = 60°C, the region homologous to plasmid pLOI3421 is underlined)

   xylB rev KO: 5′-CCC CCA CCC GGT CAG GCA GGG GAT AAC GTT TAC GCC ATT AAT GGC AGA AG TTC CGG TCT CCC TAT AGT GA-3′ ($T_{m,xylB}$ = 82°C, $T_{m,3421}$ = 60°C, the region homologous to plasmid pLOI3421 is underlined)

   xylB for check: 5′-CGG GAT AGA TCT TGG CAC CTC-3′ (T_m = 64°C)

   xylB rev check: 5′-AGG CAG GGG ATA ACG TTT AC-3′ (T_m = 61°C)

   ***CbXR* for**: 5′-CCT TAA CAT ATG TCA AGC CCA CTT TTA ACC-3′ (T_m = 66°C, *Nde*I site underlined)

   ***CbXR* rev**: 5′-GCA TCG CTC GAG GTT AAA TAA ATG TTG GAA TAT-3′ (T_m = 67°C, *Xho*I site underlined)

   As depicted, the primers used for deleting *xylB* (**xylB for KO** and **xylB rev KO**) have 50 bases homologous to the *xylB* gene (5′), followed by 20 bases (3′) homologous to the region for amplifying the FRT-flanked apramycin resistance gene in plasmid pLOI3421 (underlined region).

5. High-fidelity DNA polymerases such as Phusion have proofreading capability and should be used whenever it is important to avoid mutations during gene amplification (e.g., when it is desired to clone an exact copy of a gene). *Taq* polymerase typically provides higher yields and is the standard choice in cases where sequence fidelity is not important, such as when PCR is used to check for the presence of a gene or when it is acceptable/desired to introduce random mutations in the gene being amplified.

6. Follow the protocol supplied by the polymerase manufacturer for exact details. We often perform 2x 50 μL PCRs to increase the amount of DNA available. The reactions are combined prior to performing the gel purification.

7. Add the agarose to 1x TAE and bring the mixture to a boil. Swirl the mixture to mix, then allow to cool slightly. Pour the melted agarose mixture into a gel mold containing a sample well comb to a depth of 3–5 mm. Check for and remove any bubbles. Allow the gel to set at room temperature for 30–45 min. Carefully remove the sample well comb and transfer the gel to a gel electrophoresis box. Add 1x TAE containing 0.5 μg/mL ethidium bromide (EtBr) to the gel box to submerge the gel.
8. Follow the protocol supplied by the ligase manufacturer for exact details. The ligation reaction contains insert and vector, typically at ~3:1 molar ratio and with the vector concentration typically ~50 ng per 20 μL reaction. An example protocol using T4 DNA ligase from Invitrogen is as follows. In a microcentrifuge tube add (in order) X H_2O, 4 μL 5x ligase reaction buffer, Y μL insert, Z μL vector, and 1 μL T4 DNA ligase, where $X+Y+Z = 15$ μL. Mix by pipetting. Incubate this reaction overnight at 16°C.
9. Failure to properly clean the ligation reaction (i.e., remove salts) prior to electroporation may cause arcing in the electroporation sample. Optimal dialysis time may vary.
10. We describe preparation of electrocompetent cells, although there are other methods of making cells competent for transformation of plasmids or linear DNA fragments. To prepare electrocompetent cells, inoculate 30 mL LB with three drops of an overnight culture and grow to $OD_{600} \sim 0.35$ (about 3 h) with 250 rpm shaking at an appropriate temperature (typically 37°C unless another temperature is required). Transfer the culture to a centrifuge tube and place on ice for ~15 min. Centrifuge the tube for 10 min at 3,750 rpm and 4°C to pellet the cells. After centrifuging, decant the supernatant and wash the cell pellet using ice-cold water, making sure to break up the cell pellet. Centrifuge and wash two more times. After the last centrifugation, decant most of the supernatant and resuspend the pellet in the remaining hold-up volume of ~300 μL.
11. Follow the protocol supplied by the manufacturer for exact details. The reaction described is for a *Taq* DNA polymerase supplied by New England BioLabs (NEB, Beverly, MA). We often perform 2x 50 μL PCRs to increase the yield of DNA. The reactions are combined prior to performing the gel purification.
12. We have found that *Dpn*I does not efficiently cut the methylated plasmid in the PCR buffer; therefore we often perform a double digest, using the restriction

endonucleases *Dpn*I and *Fsp*I, to ensure that plasmid pLOI3421 is cut. To perform the double digest, the DNA must first be cleaned using a DNA clean and concentrator kit to remove the polymerase buffer and enzyme.

13. SOB medium is used instead of SOC medium because the presence of glucose would impair L-arabinose-induced Red recombinase expression from the P_{BAD} promoter (22).
14. We have achieved better transformation/recombination efficiency when we used higher DNA concentrations (e.g., > 200 ng).
15. Half of the outgrowth medium is saved to be plated the next day in case the overnight plates do not yield colonies. The outgrowth medium can sit on a bench top at room temperature overnight before plating (14).
16. To perform the triphenyltetrazolium chloride (TTC) assay, from a single colony inoculate 3 mL LB supplemented with glucose (1%) and arabinose (1%). When the culture reaches an OD_{600} ~2.0, harvest the cells and wash twice with a phosphate buffer (pH = 7.4) containing kanamycin (50 μg/mL). After allowing for residual sugars to be cleared, suspend the cells in a buffer containing arabinose (1%), kanamycin (50 μg/mL), and TTC (1%). Reduction of TTC results in red color formation and indicates constitutive arabinose utilization (arabinose catabolism genes were expressed in the presence of glucose and catabolite repression was alleviated).

References

1. Parajo, J. C., Dominguez, H., and Dominguez, J. M. (1998) Biotechnological production of xylitol. Part 1: Interest of xylitol and fundamentals of its biosynthesis. *Bioresour. Technol.* **65**, 191–201.
2. Akinterinwa, O., Khankal, R., and Cirino, P. C. (2008) Metabolic engineering for bioproduction of sugar alcohols. *Curr. Opin. Biotechnol.* **19**, 461–467.
3. Schmid, A., Dordick, J. S., Hauer, B., Kiener, A., Wubbolts, M., and Witholt, B. (2001) Industrial biocatalysis today and tomorrow. *Nature* **409**, 258–268.
4. Duetz, W. A., van Beilen, J. B., and Witholt, B. (2001) Using proteins in their natural environment: Potential and limitations of microbial whole-cell hydroxylations in applied biocatalysis. *Curr. Opin. Biotechnol.* **12**, 419–425.
5. Chin, J. W., Khankal, R., Monroe, C. A., Maranas, C. D., and Cirino, P. C. (2009) Analysis of NADPH supply during xylitol production by engineered Escherichia coli. *Biotechnol. Bioeng.* **102**, 209–220.
6. Khankal, R., Luziatelli, F., Chin, J. W., Frei, C. S., and Cirino, P. C. (2008) Comparison between Escherichia coli K-12 strains W3110 and MG1655 and wild-type E. coli B as platforms for xylitol production. *Biotechnol. Lett.* **30**, 1645–1653.
7. Cirino, P. C., Chin, J. W., and Ingram, L. O. (2006) Engineering Escherichia coli for xylitol production from glucose-xylose mixtures. *Biotechnol. Bioeng.* **95**, 1167–1176.
8. Khankal, R., Chin, J. W., and Cirino, P. C. (2008) Role of xylose transporters in xylitol production from engineered Escherichia coli. *J. Biotechnol.* **134**, 246–252.
9. Akinterinwa, O., and Cirino, P. C. (2009) Heterologous expression of D-xylulokinase from Pichia stipitis enables high levels of xylitol production by engineered Escherichia

coli growing on xylose. *Metab. Eng.* **11**, 48–55.

10. Zheng, D., Constantinidou, C., Hobman, J. L., and Minchin, S. D. (2004) Identification of the CRP regulon using in vitro and in vivo transcriptional profiling. *Nucleic Acids Res.* **32**, 5874–5893.
11. Eppler, T., and Boos, W. (1999) Glycerol-3-phosphate-mediated repression of malT in Escherichia coli does not require metabolism, depends on enzyme IIAGlc and is mediated by cAMP levels. *Mol. Microbiol.* **33**, 1221–1231.
12. Wood, B. E., Yomano, L. P., York, S. W., and Ingram, L. O. (2005) Development of industrial-medium-required elimination of the 2,3-butanediol fermentation pathway to maintain ethanol yield in an ethanologenic strain of Klebsiella oxytoca. *Biotechnol. Prog.* **21**, 1366–1372.
13. Cherepanov, P. P., and Wackernagel, W. (1995) Gene disruption in Escherichia coli: TcR and KmR cassettes with the option of Flp-catalyzed excision of the antibiotic-resistance determinant. *Gene* **158**, 9–14.
14. Datsenko, K. A., and Wanner, B. L. (2000) One-step inactivation of chromosomal genes in Escherichia coli K-12 using PCR products. *Proc. Natl. Acad. Sci. USA* **97**, 6640–6645.
15. Kang, M. H., Ni, H., and Jeffries, T. W. (2003) Molecular characterization of a gene for aldose reductase (CbXYL1) from Candida boidinii and its expression in Saccharomyces cerevisiae. *Appl. Biochem. Biotechnol.* **105–108**, 265–276.
16. Sambrook, J., and Russell, D. W. (2001) *Molecular Cloning: A Laboratory Manual.* Cold Spring Harbor Laboratory Press, Cold Spring Harbor, NY.
17. Snyder, L., and Champness, W. (2007) *Molecular Genetics of Bacteria.* ASM Press, Washington, DC.
18. Hanahan, D., Jessee, J., and Bloom, F. R. (1995) Techniques for transformations of E. coli. In *DNA Cloning 1: Core Techniques*, Vol. 1 (Glover, D. M. and Homes, B. D., Eds.), Oxford University Press, New York, pp. 1–36.
19. Luria, S. E., Adams, J. N., and Ting, R. C. (1960) Transduction of lactose-utilizing ability among strains of E. coli and S. dysenteriae and the properties of the transducing phage particles. *Virology* **12**, 348–390.
20. Greene, J. J. (2004) Host cell compatibility in protein expression. *Methods Mol. Biol.* **267**, 3–14.
21. Lipps, G. (2008) *Plasmids: Current Research and Future Trends.* Caister Academic, Norfolk.
22. Schleif, R. (2000) Regulation of the L-arabinose operon of Escherichia coli. *Trends Genet.* **16**, 559–565.

Chapter 16

Enzyme-Carrying Electrospun Nanofibers

Hongfei Jia

Abstract

Compared to other nanomaterials as supports for enzyme immobilization, nanofibers provide a promising configuration in balancing the key factors governing the catalytic performance of the immobilized enzymes including surface area-to-volume ratio, mass transfer resistance, effective loading, and the easiness to recycle. Synthetic and natural polymers can be fabricated into nanofibers via a physical process called electrospinning. The process requires only simple apparatus to operate, yet has proved to be very flexible in the selection of feedstock materials and also effective to control and manipulate the properties of the resulting nanofibers such as size and surface morphology, which are typically important parameters for enzyme immobilization supports. This chapter describes a protocol for the preparation of nanofibrous enzyme, involving the synthesis and end-group functionalization of polystyrene, production of electrospun nanofibers, and surface immobilization of enzyme via covalent attachment.

Key words: Enzyme immobilization, covalent attachment, electrospinning, nanofiber, polystyrene, α-chymotrypsin.

1. Introduction

While enzymes demonstrated promising features as molecular catalyst because of their selectivity and mild reaction conditions, low activity and stability are often the barriers for large-scale applications to compete with traditional chemical processing (1–3). Immobilization of enzyme onto solid support has been extensively studied as one of the countermeasures to overcome these challenges, which meanwhile also promote the easy separation and recycling of the catalysts from the reaction medium (4, 5). Recently growing interest has been seen on the development of nano-structured materials as enzyme immobilization supports (6–9). It is expected that, due to the match in geometric dimension, nanomaterials afford unique opportunity in pursuing

P. Wang (ed.), *Nanoscale Biocatalysis*, Methods in Molecular Biology 743,
DOI 10.1007/978-1-61779-132-1_16,

highly stable and active biocatalysts by fine-tuning the supporting materials to create the perfect microenvironment for accommodation of enzyme molecules.

Among various types of nanomaterials used for enzyme immobilization as addressed extensively in this book and other literatures, nanofibers are of special importance for minimizing mass transfer limitation and use in sensor and filter fabrications due to the interconnectivity configuration (10). Various types of polymeric nanofibers have been prepared by electrospinning, a process requiring very simple setup, essentially just a high-voltage power supply, a spinneret, and a conductive collector, although the mechanism of fiber formation in the electric field is rather complicated (11). Electronspinning is a versatile technology that fits a wide variety of synthetic and natural polymers, as well as composite materials (12, 13). Properties of the resultant nanofibers including size, surface morphology, and fiber alignment can be controlled by manipulating the feedstock materials and the spinning conditions.

Electrospinning is a 100-year-old technology; however, the research on its application in enzyme immobilization started just in this decade (14). In general, enzyme can be either attached to the surface or entrapped inside of the fibers (10), although treatments such as additional crosslinking and use of flexible spacers proved to be effective to enhance enzyme durability and stability (15, 16). This chapter describes a protocol to the surface attachment of enzyme onto polystyrene nanofibers. The procedure shown here consists of three steps: synthesis of functionalized polystyrene, electrospinning, and enzyme immobilization. **Scheme 16.1** depicts the chemical reactions involved in the synthesis and subsequent enzyme attachment. It has been reported that nanofibers carrying a model enzyme α-chymotrypsin retained 65% of their aqueous activity and were up to thousands of times more active than free enzyme in organic solvents (17). This is mostly due to the fact that surface-attached enzymes are more accessible to reaction substrates and thus afford less diffusional limitation and higher catalytic efficiency.

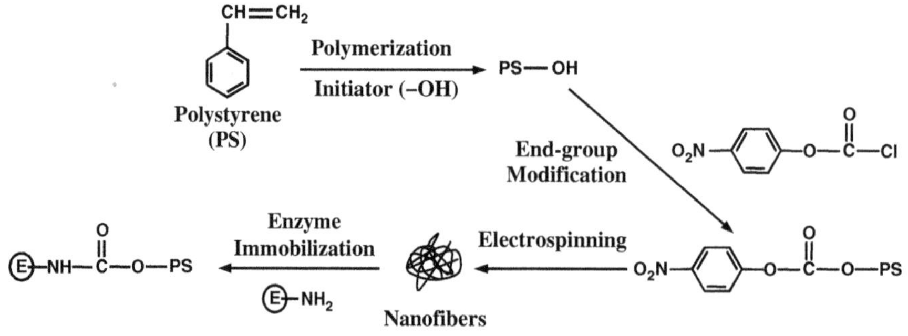

Scheme 16.1. Chemical reactions for polystyrene synthesis and enzyme attachment (*see* **Note 1**).

2. Materials

2.1. Preparation of Polystyrene (PS) with Functionalized End Groups

1. Styrene (EMD Chemicals): monomer.
2. 2,2′-Azobis[2-methyl-*N*-(2-hydroxyethyl)propionamide] (VA-086, Wako Chemicals, USA): initiator.
3. 4-Nitrophenyl chloroformate (NPC, Aldrich), 4-dimethylaminopyridine (DMAP, Sigma).
4. Toluene, dimethylchloride *N,N*-dimethylformamide (DMF), and methanol.
5. Purging gas: N_2.
6. Glass vials (20 mL, scintillation).
7. Centrifuge.
8. Magnetic stir plate and stir bars.

2.2. Electrospinning

1. LiCl (Aldrich).
2. Solvents: methyl ethyl ketone (MEK), DMF.
3. High-voltage DC power supply (Gamma High Voltage Research, ES30P-5 W).
4. Syringe pump.
5. Syringes and flat-headed syringe needles.
6. Glass microscope slides (25 × 75 mm).
7. Stainless steel meshes (type 304 woven wire cloth, 24 × 24, 0.014 in. wire diameter, from Mcmaster-Carr Supply Company, Aurora, OH).
8. Aluminum foils (0.016 mm thickness).

2.3. Enzyme Immobilization

1. α-Chymotrypsin (CT, Sigma).
2. Sodium borate buffer solution (pH 8.2, 10 mM).
3. Sodium borate buffer solution (pH 7.5, 10 mM).
4. *n*-Succinyl-ala-ala-pro-phe *p*-nitroanilide (SAAPPN, Sigma): substrate for enzyme activity assay.
5. *p*-Nitroaniline (Sigma): calibration standard for enzyme activity assay.
6. 4-Methylumbelliferyl *p*-trimethylammonium cinnamate chloride (MUTMAC, Sigma): use in active site titration to determine active enzyme loading on nanofibers.
7. UV–visible spectrophotometer and luminescence spectrometer.
8. 0.22 μm PTFE syringe filters.

3. Methods

3.1. Synthesis of Polystyrene

1. Add 0.1 g of VA-086 and 4 mL of DMF into a 20-mL glass vial and mix to dissolve all the solids.
2. Mix the solution with 5 mL of styrene and 1 mL of toluene in the glass vial.
3. Purge the glass vial with N_2.
4. Incubate the reaction mixture in water bath at 72°C for 24 h.
5. Pour the reaction mixture into a beaker with 50 mL of methanol to precipitate the product PS.
6. Decant the solvents and wash PS with 25 mL of methanol for at least three times.
7. Evaporate the solvents by purging N_2 to obtain dry PS product.
8. Analyze molecular weight of the PS product using gel permeation chromatography (*see* **Note 2**).

3.2. Modification of Polystyrene End Groups

1. Dissolve 0.5 g of PS and 12.2 g of DMAP with 8 mL of toluene in a 20-mL scintillation vial.
2. Cool the solution to 4°C in a refrigerator.
3. Initiate the reaction by adding 2 mL of 0.01 M NPC (in methylene chloride) under magnetic stirring.
4. Seal the vial and continue to stir the solution for 5 h in the refrigerator.
5. Centrifuge the reaction mixture to precipitate the byproduct of 4-dimethylaminopyridine hydrochloride.
6. Decant the supernatant into 50 mL of methanol to precipitate the nitrophenyl-functionalized PS (PS-NPh).
7. Wash the product with 25 mL of methanol three times.
8. Dry the product by purging N_2 and store at 4°C in a refrigerator.

3.3. Preparation of Nanofibers via Electrospinning

1. Prepare PS-NPh solution in a mixture of MEK and DMF (1:1 v/v) containing 0.5 wt% of LiCl (*see* **Notes 3** and **4**).
2. Set up the electrospinning system as illustrated in **Scheme 16.2**. Connect the jet (flat-headed syringe needle) and the collecting substrates to the high voltage and the ground electrodes of the power supply, respectively.
3. Load polymer solution into the syringe and spin fibers at an electric field strength of 0.75 kV/cm.

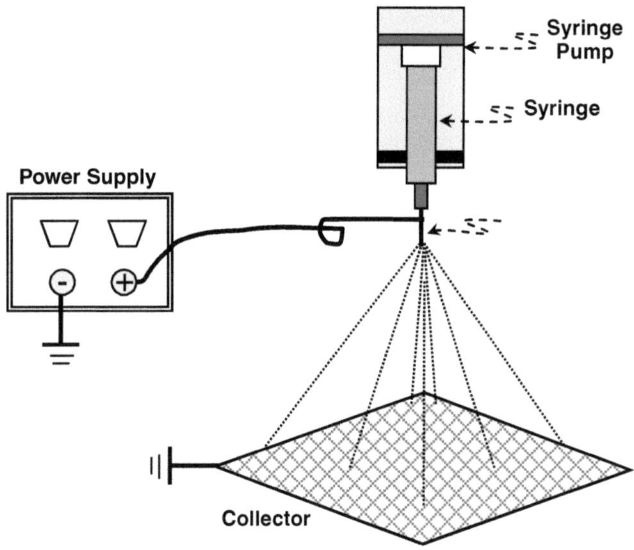

Scheme 16.2. Illustration of the experimental setup for electrospinning.

4. Collect fibers on stainless steel meshes for enzyme immobilization (*see* **Note 5**).
5. Measure the weight of the meshes before and after collection to determine the amount of fibers being deposited on the substrate.

3.4. Enzyme Immobilization

1. Prepare 15 mL of 5 mg/mL CT solution with pH 8.2 borate buffer in a 20-mL glass vial.
2. Cut the fiber-carrying stainless steel meshes into small pieces (1 × 2 cm) and immerse them in the enzyme solution.
3. Slightly shake the vial at room temperature for 36 h.
4. Rinse the meshes with fresh pH 8.2 borate buffer until no absorbance at 280 nm is detected on a UV–visible spectrophotometer.
5. Dry the fibers by purging N_2.

3.5. Enzyme Loading Assay

1. Prepare MUTMAC solution at a concentration of 0.025 mg/mL with pH 7.5 borate buffer (*see* **Note 6**).
2. Add roughly 1 mg of nanofibrous enzyme into 3 mL of MUTMAC solution.
3. Allow them to react for 2 min.
4. Filter the solution using 0.22 μm PTFE syringe filters.
5. Measure the fluorescence of the filtrate at an excitation wavelength of 360 nm and an emission wavelength of 450 nm.
6. Conduct parallel measurements using fibers that have no enzyme attached.

7. Calculate the difference in fluorescence between fibers with and without enzyme and then fit it into the calibration curve to estimate the active enzyme loading.

3.6. Enzyme Activity Assay

1. Prepare 0.8 mM of SAAPPN solution with pH 8.2 borate buffer (*see* **Note 7**).
2. Prepare five 20-mL glass vials. To each vial, add 4 mL of SAAPPN and ~0.1 mg of nanofibrous enzyme.
3. Stop the reactions at different time intervals (2–20 min) by filtering out the fibers with 0.22 μm PTFE syringe filters.
4. Measure the absorbance of filtrates at 410 nm and record the change of absorbance vs. time.
5. Convert the rate of absorbance change into product generation rates based on a calibration curve (*see* **Note 8**).
6. Calculate specific activity based on active enzyme loadings and product generation rates.

4. Notes

1. The chemical routes for polymer modification and enzyme attachment are adopted from a previous literature (18).
2. Following the protocol, the molecular weight of the resultant PS should be ~200 kDa.
3. LiCl is introduced into the solution to suppress the formation of beaded fibers. SEM images in **Fig. 16.1** show the difference in morphology of electrospun nanofibers with and

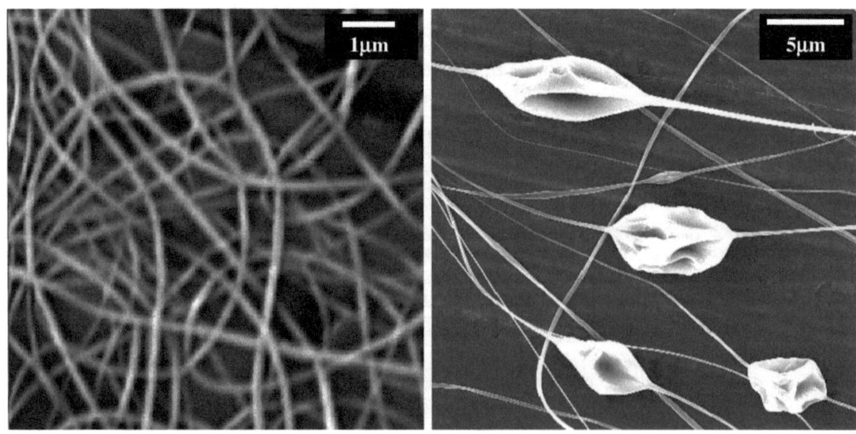

Fig. 16.1. Polystyrene fibers prepared with (*left*) and without (*right*) LiCl PS concentration: 8.0 wt%; LiCl concentration: 0.5 wt%.

without the salt. For easy preparation, first dissolve LiCl in DMF at a higher concentration and then dilute with MEK to the desired concentration.

4. Concentration of the polymer solution directly affects the size of the fibers – the higher the concentration, the larger the fiber diameters. Recommended range of PS concentration in the mixed solvents for this protocol is 5–20 wt%.
5. Cover the collecting area (ground) with aluminum foils. Always use glass slides to collect fibers during a spinning process to verify fiber formation with optical microscope.
6. During the active site titration, MUCMAC specifically binds to the active site of a-CT (19). Prepare a stock solution at a higher concentration (1 mg/mL) and dilute it before each test. Keep the stock solution in dark and discard it if older than a few hours.
7. Make a stock solution of SAAPPN in DMF at 40 mM and dilute it to 0.8 mM for each test. Keep the stock solution in dark at –20°C.
8. Use *p*-nitroaniline to establish a calibration curve prior to activity assay.

Acknowledgments

The protocol was originally developed at Prof. Ping Wang's lab in the Department of Chemical Engineering at the University of Akron, with financial support from National Science Foundation NER program (Award #0103232).

References

1. Rozzell, J. D. (1999) Commercial scale biocatalysis: Myths and realities. *Bioorg. Med. Chem.* 7(10), 2253–2261.
2. Wandrey, C., Liese, A., and Kihumbu, D. (2000) Industrial biocatalysis: Past, present, and future. *Org. Process Res. Dev.* **4**(4), 286–290.
3. Schmid, A., et al. (2001) Industrial biocatalysis today and tomorrow. *Nature* **409**(6817), 11.
4. Kennedy, J. F., Melo, E. H. M., and Jumel, K. (1990) Immobilized enzymes and cells. *Chem. Eng. Prog.* **86**(7), 81–89.
5. Tischer, W., and Wedekind, F. (1999) Immobilized enzymes: Methods and applications. *Top. Curr. Chem.* **200**(Biocatalysis: From Discovery to Application), 95–126.
6. Kim, J., Grate, J. W., and Wang, P. (2009) Nanobiocatalysis and its potential applications. *Trends Biotechnol.* **26**(11), 8.
7. Kim, J., Grate, J. W., and Wang, P. (2006) Nanostructures for enzyme stabilization. *Chem. Eng. Sci.* **61**(3), 1017–1026.
8. Wang, P. (2006) Nanoscale biocatalyst systems. *Curr. Opin. Biotechnol.* **17**(6), 574–579.
9. Wang, P. (2009) Multi-scale features in recent development of enzymic biocatalyst systems. *Appl. Biochem. Biotechnol.* **152**(2), 343–352.

10. Wang, Z.-G., et al. (2009) Enzyme immobilization on electrospun polymer nanofibers: An overview. *J. Mol. Catal. B: Enzym.* **56**(4), 7.
11. Xie, J., Li, X., and Xia, Y. (2008) Putting electrospun nanofibers to work for biomedical research. *Macromol. Rapid Commun.* **29**(22), 18.
12. Li, D., and Xia, Y. N. (2004) Electrospinning of nanofibers: Reinventing the wheel? *Adv. Mater.* **16**(14), 1151–1170.
13. Reneker, D. H., and Chun, I. (1996) Nanometer diameter fibers of polymer, produced by electrospinning. *Nanotechnology* 7(3), 216–223.
14. Kim, B. C., et al. (2008) Enzyme-nanofiber composites for biocatalysis applications. *ACS Symp. Ser.* **986**(Biomolecular Catalysis), 9.
15. Wang, Y., and Hsieh, Y.-L. (2004) Enzyme immobilization to ultra-fine cellulose fibers via amphiphilic polyethylene glycol spacers. *J. Polym. Sci., Part A: Polym. Chem.* **42**(17), 4289–4299.
16. Byoung Chan, K., et al. (2005) Preparation of biocatalytic nanofibres with high activity and stability via enzyme aggregate coating on polymer nanofibres. *Nanotechnology* **16**(7), S382.
17. Jia, H., et al. (2002) Enzyme-carrying polymeric nanofibers prepared via electrospinning for use as unique biocatalysts. *Biotechnol. Prog.* **18**(5), 1027–1032.
18. Zacchigna, M., et al. (1998) Properties of methoxy(polyethylene glycol)-lipase from Candida rugosa in organic solvents. *Farmaco* **53**(12), 6.
19. Gabel, D. (1974) Active site titration of immobilized chymotrypsin with a fluorogenic reagent. *FEBS Lett.* **49**(2), 2.

Chapter 17

Uniform Lab-Scale Biocatalytic Nanoporous Latex Coatings for Reactive Microorganisms

Jimmy L. Gosse and Michael C. Flickinger

Abstract

This chapter describes a method for generating uniform lab-scale biocatalytic nanoporous latex coatings. Nearly everything we come into contact with on a daily basis has been coated with some polymer material. High-speed waterborne polymer coating and ink-jet printing techniques are mature technologies. Methods for immobilizing microorganisms in lab-scale waterborne latex biocatalytic coatings draw on existing coating technologies for generating precision industrial paint and paper coatings and would therefore be amenable to scale up in future applications. An inherent problem for many lab-scale techniques is coating uniformity. The method described here has been developed to dramatically increase the uniformity of multiple individual small surface area coatings derived from a single coating template by minimizing edge effects due to emulsion drying adjacent to the edge of the mask.

Key words: Latex immobilization, coating bioreactor, bacteria, microbial, biocatalysis, nanoporous coatings, embedded, synthetic biofilms.

1. Introduction

Microbial biotechnology using living microbes as biocatalysts is as old as fermentation and the leavening of bread. Industrially produced microbial products are worth hundreds of billions of dollars in the USA alone (1). Many more microbial processes would be industrially feasible; however, the rates of reaction are low and process intensities in traditional suspended cell reactor geometries are not high enough for commercialization. Natural and synthetic (polymer and microbes) microbial biofilms have been shown to have increased stability, process intensity, long-reactive half-lives, and decreased sensitivity to system stresses

P. Wang (ed.), *Nanoscale Biocatalysis*, Methods in Molecular Biology 743,
DOI 10.1007/978-1-61779-132-1_17,

over comparable suspended cell systems (2–4). A diverse number of microorganisms have been used as substrate-limited non-growth immobilized biocatalysts in thin (5–75 μm thick) adhesive nanoporous latex polymer coatings. Among the organisms that have been immobilized are a strict aerobe, *Gluconobacter oxydans* (5); facultative anaerobes, numerous *Escherichia coli* strains (6–13); an anaerobic phototroph *Rhodopseudomonas palustris* (14, 15); as well as a thermophilic anaerobic marine bacterium, *Thermotoga maritima* (16). Each of these organisms has unique constraints and requirements that must be accounted for during the coating and drying process in order to generate a metabolically active immobilized biocatalyst. Edge effects are a problem common to all of these small surface area coating applications due to the production of single and multilayer uniform coatings using a single coating template. Raised edges resulting from polymer flow during drying lead to coating flaws when top coated with subsequent layers of nanoporous polymer (17). Prior template designs generated many individual patch or strip coatings; however, each was an individual coating separated from the others by a mask (similar to letters in a stencil). As a result, the perimeter of every coating contained numerous edge effects from latex emulsion drying in contact with the pressure-sensitive vinyl mask. The strip method described here limits the edge effects on individual coatings to the ends of the strip which for a 1 cm wide 10 cm long strip is only 1/11 of the total perimeter (compared to 100% of the perimeter in former template designs).

The template design described in this chapter is used to generate multiple uniform strip coatings on a polyester substrate and includes all the basic components that can be adapted to create new geometries for specific applications (**Fig. 17.1**). The coating template consists of a support, substrate, adhesive layer, and a mask. The substrate is the surface to which the coating will be applied. In most cases the coating will adhere to the substrate following drying and as a result the substrate and coating will be inseparable throughout an experimental procedure. For this reason choosing a coating substrate which is compatible with all experimental constraints is important (i.e., flexible, conductive, appropriate surface energy for polymer adhesion; clear for photosynthetic work; thermostable for high-temperature work). Common laboratory substrates include polyester, plastic, glass, stainless steel, paper, photographic paper, sintered glass, fibers, and a number of commercially available filters. During the coating process it is critical to ensure that the substrate remains in place and is level. An adhesive layer is applied to a glass support to fix the substrate in place. Although the most common flat support is glass, other supports such as metal shim stock have been used. The final component is the pressure-sensitive vinyl mask which is used to

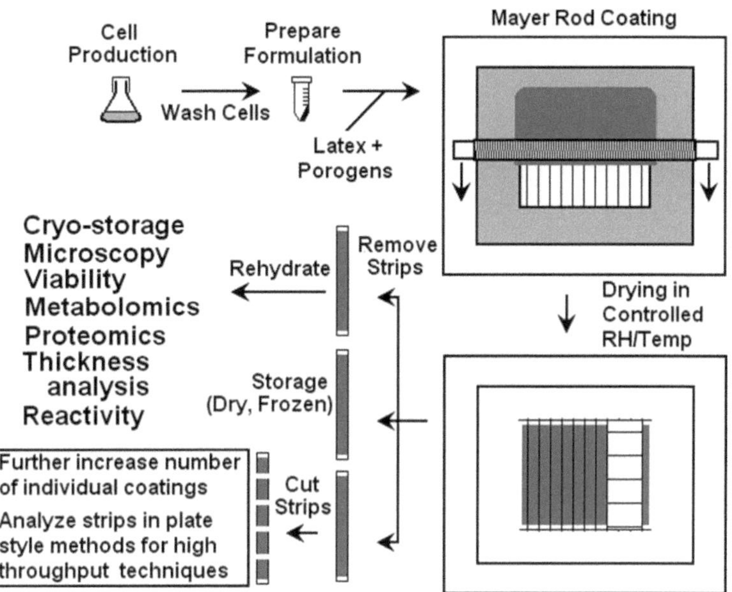

Fig. 17.1. Preparation of multiple uniform biocatalytic latex strip coatings of reactive microbes on a polyester substrate. This method minimizes non-uniform strip edge effects such as raised edges adjacent to the mask. All steps are non-aseptic after cell production when assays are performed under nongrowth conditions. The method can be modified with a double mask to generate porous polymer multi-layer or top-coated strips.

ensure that all of the individual coatings have the same surface area. An adhesive (one-sided) vinyl mask is cut or punched to the appropriate dimensions and applied directly to the substrate using a rubber roller. **Figure 17.2** shows each of the components and how they are assembled to form the complete template.

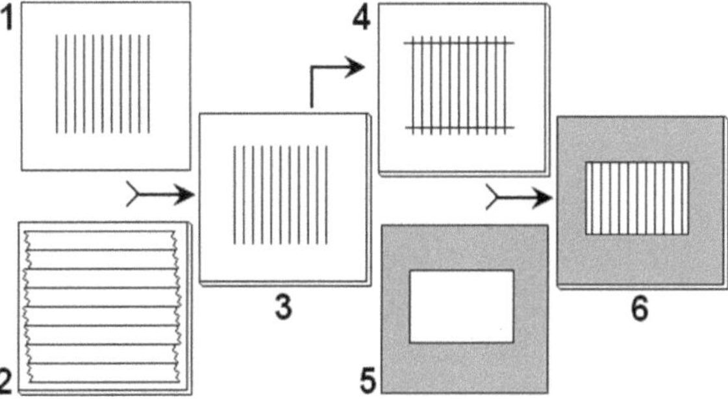

Fig. 17.2. Coating template assembly. (*1*) Polyester substrate cut with parallel vertical lines. (*2*) Glass support covered with double-sided adhesive tape. (*3*) Substrate affixed to support. (*4*) Horizontal cuts made for final strip lengths. (*5*) Mask cut. (*6*) Mask placed on substrate centered between the horizontal cuts and extending slightly beyond either side of vertical cuts.

2. Materials

2.1. Template Assembly

1. Marker
2. Razor blade
3. Ruler (30 cm)
4. 20 × 20″ glass plate support
5. Double-sided removable tape (3 M, St. Paul, MN, USA)
6. Transparent vinyl shelf covering (84 ± 4 μm thick Con-Tact, Stamford, CT, USA)
7. Polyester sheet substrate (DuPont Melinex 454, Tekra Corp, NJ, USA)
8. 4″ hard rubber roller (Dick Blick Art Materials, Galesburg, IL, USA)

2.2. Formulation

1. Spatula (Fisher, handi-hold microspatula)
2. Rhoplex™ SF-012 or comparable acrylate/vinyl acetate latex binder which is biocide free, pH ~7, and adhesive to polyester after rehydration (Rhoplex™ SF-012, Rohm and Haas Co., Philadelphia, PA, USA)
3. Wet cell pellet (*see* **Note 7**)
4. Media for washing cells (*see* **Note 8**)
5. Glycerol (Fisher, neat)
6. Sucrose (Fisher, 0.58 g/ml)
7. 50 ml conical tubes (Fisher)
8. 500 ml centrifuge bottles (Fisher)

2.3. Applying the Formulation and Drying

1. #26 1/4″ Diameter wet film applicator rod – Mayer rod (Gardco, Paul N. Gardner Co., Pompano Beach, FL, USA)
2. Controlled humidity and temperature chamber (Model 5532, Electro-Tech Systems Inc, Glenside, PA, USA)
3. Surface level (Small Parts Inc., Miramar, FL, USA)
4. Straight serrated forceps (Fisher)
5. Razor blade
6. Tooth brush
7. Ethanol (70%)

3. Method

The strip template design chosen for this chapter includes the basic components which can be altered as needed for a specific application. Template design is completely dependent on the desired surface area, end use of the coating, as well as the geometry of the vessel the coating will be placed into (micro-well, tube, petri dish, etc.). The following template produces 10 coatings that are 1 cm wide and 10 cm long with an approximate thickness of 60 μm. Thickness determinations are routinely performed on rehydrated coatings using a bench top digital micrometer (Mitutoyo, Kawasaki, Japan); however, microscopy (optical, confocal, SEM, and AFM) and profilometry have also been used. Dry coating thickness is a function of the percent solids in the coating emulsion and the metering of the latex onto the substrate. In this method, metering is by wire-wound Mayer rod (18). If more strips are needed each one can be cut into half or into more pieces to generate the number of coatings needed from a single template (**Fig. 17.1**). The high catalytic potential of these strips is a result of the concentration of the microorganisms in the adhesive coating. Starting with a culture containing 10^9 CFU/ml, this method will generate a formulation with approximately 10^{11} CFU/ml and a coating with 8×10^{13} CFU/m^2 of coating surface area.

3.1. Template Assembly

1. Begin by cutting a 17 × 17 cm transparent vinyl sheet from the roll of pressure-sensitive transparent vinyl shelf covering (*see* **Notes 1** and **2**).
2. Measuring from the top of the 17 × 17 cm sheet, draw lines horizontally at 3.5 and 13.5 cm.
3. Measuring from the left edge draw vertical lines at 3 and 14 cm. Cut out and discard the resulting square in the center of the vinyl using a razor blade and the ruler as a straight edge.
4. Put aside the vinyl mask (outer square) and cut a polyester sheet to 17 × 17 cm.
5. Label the top edge with TP and the left edge with an L.
6. Draw a dashed line across the polyester 3 and 14 cm down from the top edge.
7. Along the top dashed line, starting from the left side, make small marks at 3 and 3.5 cm followed with a mark every 1–13.5 cm. Finish with a mark at 14 cm. Do the same for the bottom dashed line.
8. Using the ruler as a straight edge cut the polyester vertically (top to bottom) at each mark from 3.5 to 12.5 cm. *Do not* completely cut from the top to bottom of the sheet. Extend the cuts from 0.5 to 1 cm above and below the dashed lines.
9. The end result should be 11 parallel cuts ~12 cm in length, centered about the dashed lines (*see* **Note 3**).
10. On a clean surface place the polyester face down (the L should be backward and on the right side). Place the glass

plate on top of the polyester which will serve as a guide for cutting each length of tape.

11. Carefully put down a piece of removable double-sided tape starting at the bottom of the glass plate. Use your index finger (clean and dry) to smooth out any wrinkles or bubbles in the tape prior to laying down the next row (*see* **Note 4**).
12. Work toward the top, adding rows of tape ensuring each new row is as close as possible to the previous row without overlapping.
13. Once the glass plate has sufficient tape to cover the entire polyester sheet, pick up the glass plate and place it tape-side down on the polyester. Be sure that the polyester is completely covered and that the tape is perpendicular to the cuts.
14. Turn over the glass plate (polyester is now stuck to the double-sided tape) and use the rubber roller to carefully press the polyester down and remove any trapped air pockets (*see* **Note 5**). Trim away any exposed double-sided tape.
15. Using the razor blade and the ruler as a straight edge, cut the polyester along the horizontal dashed lines extending 0.5–1 cm beyond the first and last of the series of vertical cuts.
16. Retrieve the vinyl mask and separate the paper backing from the top edge down approximately 2 cm. Cut away the paper.
17. Center the vinyl on the template with the top edge of the square approximately 0.5 cm below the top dashed line, so that there is approximately 0.5 cm on either side of the outside vertical cuts. Use a roller to firmly press down the top of the vinyl mask.
18. Remove the remaining paper backing and carefully orient the mask while slowly placing it on the polyester (*see* **Note 6**). Use the roller once more to firmly press the remaining area of the mask onto the polyester.

3.2. Formulation

1. The cells to be immobilized should be grown to either a desired optical density or a predetermined point in the growth curve (*see* **Note** 7, typically mid-log phase).
2. Centrifuge the culture. Discard the supernatant and resuspend pellet in <40 ml of media (*see* **Note 8**) and transfer to preweighed 50 ml conical tube.
3. Centrifuge a second time and discard the supernatant. Wait for 1 min and then remove any excess media with a pipette.

4. Weight the conical tube and calculate the wet cell weight. Using the following ratio calculate the amount of glycerol, sucrose, and latex for preparing the formulation (*see* **Note 9**):

   1.2 g wet cell weight

   1 ml latex emulsion

   350 μl sucrose (0.58 g/ml)

   150 μl glycerol (neat)
5. Add the glycerol and sucrose first. Mix thoroughly with a spatula breaking up any cell clumps while limiting the introduction of bubbles.
6. After thoroughly mixing add the latex. Mix the formulation with a spatula until homogenous (*see* **Note 10**).

3.3. Making the Coating and Drying

1. Set an environmental chamber to 30°C and 60% relative humidity.
2. Using the surface level ensure the coating template is level.
3. Firmly roll the template a last time and wipe it clean. If necessary tape the template to the bottom of the chamber to ensure it does not slide when the coating is drawn.
4. With a 1 ml pipette add the formulation across the top of the mask. The formulation should form a uniform single bead on the mask just above the exposed polyester extending slightly beyond either side of the opening (*see* **Note 11**).
5. Place the wet film applicator rod on the mask above the formulation. Slowly pull the rod into contact with the formulation and pause allowing the formulation to come into contact with the rod along the entire length of the formulation bead.
6. In one fluid motion draw the formulation across the entire template. *Do not* roll the rod. Hold it firm so that it slides across the mask as the coating is drawn leaving the excess latex formulation on the glass plate.
7. Immediately remove the rod from the chamber and thoroughly clean off the formulation with warm water. Use the toothbrush to scrub any residual formulation.
8. Cover the entire length of the rod with 70% ethanol and allow to dry.
9. The coating is dried for 1 h at 60% relative humidity and 30°C.
10. After drying remove the mask.

11. Carefully lift the corner of the first strip using the razor blade. With the forceps pull the strip off the template. Repeat for remaining strips (*see* **Note 12**).

4. Notes

1. It is important to wear gloves whenever handling the polyester and mask. Finger prints can disrupt the final coating.
2. The polyester and vinyl mask should not be wiped with any concentration of ethanol (*see* **Note 13**).
3. It is critical that the parallel cuts are precise because they will define the surface area of the final coating.
4. Do not wear gloves during this step. The strips of tape should be as close together as possible without overlapping. If there is a misalignment resulting in a gap, do not start the next row, immediately remove the misaligned piece of tape and either reapply it or get a new piece.
5. Make sure the polyester is clean and free of any debris before affixing it to the glass plate. Trapped air can be removed by firmly pressing an edge of the roller to the bubble then slowly working it toward the closest tape edge.
6. The vinyl mask is easily stretched so be careful not to distort its final shape. When the mask makes full contact with the substrate immediately make sure the top is correct and the vinyl is wrinkle free along the top edge of the mask. Then proceed to slowly tap down the inner edges of the mask from the top to the bottom. If a wrinkle is encountered gently press it out working from the center of the mask out. If it cannot be worked out try to lift the mask off the affected edge and replace. In the worst case remove the mask and make a new one.
7. This method is not for a specific microorganism or latex emulsion. The time at which cells are harvested must be a balance between the condition the cells must be in for the intended assay that will be performed, with the surface area of the final coating and the volume of culture needed to generate biomass needed to make the latex formulation.
8. The media utilized is dependent on the final application. A defined media lacking a nitrogen or carbon substrate is often used.
9. For example, the following calculations would be for a wet cell pellet of 0.6 g:

$0.6/1.2 = 0.5$ mL latex

$0.5 \times 350 = 175$ μL sucrose

$0.5 \times 150 = 75$ μL glycerol

10. It is critical to completely break up the cell pellet and make sure the formulation is homogenous. If the formulation appears to become gelatinous or sticky it is likely the cells have lysed. If the problem persists it may be necessary to alter the time of cell harvest, media conditions, and/or the latex formulation, or change the microbial strain to one that does not lyse when mixed with latex formulation.
11. Any bubbles in the formulation should be removed using the pipette prior to drawing the coating.
12. The strips are removed from the template and immediately immersed in the media being used for assessing the biocatalytic activity; however, many different assays can be carried out in addition to storage or dry shipment to an alternative location (**Fig. 17.1**).
13. Some latex emulsion formulations may not evenly wet the mask or polyester and will instead pool resulting in a defective coating. In some cases this problem can be prevented in future coatings by wiping the template component(s) with 70% ethanol to alter the surface energy. Substrate surface energy can be measured by water droplet contact angle.

References

1. Demain, A. L. (2000) Small bugs, big business: The economic power of the microbe. *Biotechnol. Adv.* **18**(6), 499–514.
2. Mazid, M. A. (1993) Biocatalysis and immobilized enzyme/cell bioreactors. *Nat. Biotechnol.* **11**(6), 690–695.
3. Flickinger, M. C., Schottel, J. L., Bond, D. R., Aksan, A., and Scriven, L. E. (2007) Painting and printing living bacteria: Engineering nanoporous biocatalytic coatings to preserve microbial viability and intensify reactivity. *Biotechnol. Progr.* **23**(1), 2–17.
4. Flickinger, M. C., Lyngberg, O. K., Freeman, E. A., Anderson, C. R., and Laudon, M. C. (2009) Formulation of reactive nanostructured adhesive microbial inkjet inks for miniature biosensors and biocatalysis. In *Nanotechnology Applications in Coatings* (Fernando, R. H. and Sung, L. P., Eds.), ACS Symposium Series 1008, American Chemical Society, Washington, DC, pp. 156–187.
5. Fidaleo, M., Charaniya, S., Solheid, C., Diel, U., Laudon, M., Ge, H., Scriven, L. E., and Flickinger, M. C. (2006) A model system for increasing the intensity of whole-cell biocatalysis: Investigation of the rate of oxidation of D-sorbitol to L-sorbose by thin bi-layer latex coatings of non-growing *Gluconobacter oxydans*. *Biotechnol. Bioeng.* **95**(3), 446–458.
6. Lyngberg, O. K., Ng, C. P., Thiagarajan, V., Scriven, L. E., and Flickinger, M. C. (2001) Engineering the microstructure and permeability of thin multilayer latex biocatalytic coatings containing *E. coli*. *Biotechnol. Progr.* **17**(6), 1169–1179.
7. Lyngberg, O. K., Stemke, D. J., Schottel, J. L., and Flickinger, M. C. (1999) A single-use luciferase-based mercury biosensor using *Escherichia coli* HB101 immobilized in a latex copolymer film. *J. Ind. Microbiol. Biotechnol.* **23**(1), 668–676.
8. Lyngberg, O. K., Thiagarajan, V., Stemke, D. J., Schottel, J. L., Scriven, L. E., and

Flickinger, M. C. (1999) A patch coating method for preparing biocatalytic films of *Escherichia coli*. *Biotechnol. Bioeng.* **62**(1), 44–55.

9. Mota, M., Yelshin, A., Fidaleo, M., and Flickinger, M. C. (2007) Modelling diffusivity in porous polymeric membranes with an intermediate layer containing microbial cells. *Biochem. Eng. J.* **37**(3), 285–293.
10. Schottel, J. L., Orwin, P. M., Anderson, C. R., and Flickinger, M. C. (2008) Spatial expression of a mercury-inducible green fluorescent protein within a nanoporous latex-based biosensor coating. *J. Ind. Microbiol. Biotechnol.* **35**(4), 283–290.
11. Swope, K. L., and Flickinger, M. C. (1996) Activation and regeneration of whole cell biocatalysts: Initial and periodic induction behavior in starved *Escherichia coli* after immobilization in thin synthetic films. *Biotechnol. Bioeng.* **51**(3), 360–370.
12. Thiagarajan, V., Swope, K. L., and Flickinger, M. C. (1996) Investigation of oxygen consumption by *E. coli* immobilized in a synthetic biofilm using a thin film plug reactor (TFPR). In *Progress in Biotechnology, Vol. 11; Immobilized Cells: Basics and Applications* (Wijffels, R. H., Buitelar, R. M., Bucke, C., and Tramper, J., Eds.), Elsevier Science, Inc. Elsevier Science Publishers Ltd., Amsterdam, pp. 304–312.
13. Thiagarajan, V. S., Huang, Z. S., Scriven, L. E., Schottel, J. L., and Flickinger, M. C. (1999) Microstructure of a biocatalytic latex coating containing viable *Escherichia coli* cells. *J. Colloid Interface Sci.* **215**(2), 244–257.
14. Gosse, J. L., Engel, B. J., Rey, F. E., Harwood, C. S., Scriven, L. E., and Flickinger, M. C. (2007) Hydrogen production by photoreactive nanoporous latex coatings of nongrowing *Rhodopseudomonas palustris* CGA009. *Biotechnol. Progr.* **23**(1), 124–130.
15. Gosse, J. L., Engel, B. J., Hui, J. C. H., Harwood, C. S., and Flickinger, M. C. (2010) Progress toward a biomimetic leaf: 4,000 h of hydrogen production by coating-stabilized nongrowing photosynthetic *Rhodopseudomonas palutris. Biotechnol. Progr.* **26**(4), 907–918.
16. Lyngberg, O. K., Solheid, C., Charaniya, S., Ma, Y., Thiagarajan, V., Scriven, L. E., and Flickinger, M. C. (2005) Permeability and reactivity of *Thermotoga maritima* in latex bimodal blend coatings at 80°C: A model high temperature biocatalytic coating. *Extremophiles* **9**(3), 197–207.
17. Flickinger, M. C., Fidaleo, M., Gosse, J. L., Polzin, K., Charaniya, S., Solheid, C., Lynberg, O. K., Laudon, M., Ge, H., Schottel, J. L., Bond, D. R., Aksan, A., and Scriven, L. E. (2009) Engineering nanoporous bioactive smart coatings containing microorganisms: Fundamentals and emerging applications. In *Smart Coatings II* (Provider, T. and Baghdachi, J., Eds.), ACS Symposium Series 1002, American Chemical Society, Washington, DC, pp. 52–94.
18. MacLeod, D. M. (2006) Wire-wound rod coating. In *Coating Technology Handbook*, 3rd Edn. (Tracton, A. A., Ed.), Taylor and Francis, Boca Raton, FL, pp. 18.1–18.8.

Chapter 18

Entrapment of Enzymes in Nanoporous Sol–Gels

Andreas Buthe

Abstract

The prerequisite for many successful enzyme-based biotechnologies is the preparation of highly stable and active biocatalysts, which can be achieved effectively by immobilization. This chapter introduces the immobilization of enzymes by entrapment in nanoporous silica particles made in a sol–gel process. These easily tailorable materials have been proven very beneficial for a broad variety of applications of biocatalysts. Besides the spatial confinement in silica sol–gels, another advantage is given by the easy possibility of fine-tuning the physicochemical properties of the matrix itself to provide the ideal environment for the reaction and the biocatalyst. Preparation details are demonstrated using the process of immobilizing a lipase in a sol–gel matrix, which is chemically modified by using methyl-, ethyl-, propyl-, and *i*-butyltrimethoxysilane. The transesterification of canola oil with methanol is used as a model reaction.

Key words: Enzyme immobilization, entrapment, silanes, sol–gel process, nanoporous silica glass, alkoxysilanes, lipase, transesterification, biodiesel.

1. Introduction

A broad variety of methods are available for the immobilization of biocatalysts. Since every method encounters some pros and cons, the choice of the best procedure needs to be balanced with respect to the characteristics of desired enzyme-catalyzed reaction (1, 2). When one is challenged with selecting the best suited immobilization method for a certain application, it could be a daunting task to screen all methods available as little literature offers clear recommendations. However, things can be easier if an “adjustable” material is used, whose physical and chemical properties can be fine-tuned over a broad range. Typically the solid supports and precursors used for immobilization are right off the shelf, which makes it difficult to introduce customized

P. Wang (ed.), *Nanoscale Biocatalysis*, Methods in Molecular Biology 743,
DOI 10.1007/978-1-61779-132-1_18,

changes. But such modifications are sometimes desirable since they may have a crucial influence on the efficiency of the immobilized biocatalyst (3–6). It is well known that the surrounding microenvironment affects the enzyme structure and thus the catalytic performance to a great extent as well as the manner in which substrate and product can travel to and from the active site of the enzyme (2, 3). In this respect the immobilization of biomolecules in silicate glass formed through a sol–gel process is a highly suited method. Sol–gel process denotes the transition of liquid "sol" into a solid "gel" and the resulting gel is often referred to as a sol–gel. Typically the sol phase comprises an alkoxide precursor like tetramethoxysilane that is hydrolyzed in the presence of an acid or a base catalyst yielding silanol groups (Si–OH) that are capable of undergoing a condensation to form siloxanes (Si–O–Si). Hydrolysis and condensation are accompanied by the formation of colloidal particles as well as the release of alcohol (**Fig. 18.1**). Proceeding polycondensation causes the formation of a gel in the form of a rigid interconnected porous network (7, 8). If biomolecules like enzymes are present during the gelation, they become tightly encaged in the formed network. Recent developments in the sol–gel science have resulted in a highly flexible technology platform for the generation of tailorable nanoporous silica glasses, suitable for the entrapment of diverse bioactive substances (8). Sol–gels are easily producible and have a good thermal and chemical stability. Due to the open pore structure and controllable pore sizes, nanoporous materials like sol–gels offer a large surface area in which mass transfer limitations are less severe than is observed for other immobilization methods like the entrapment in polymers (8).

Basically there are two ways to immobilize biomolecules in sol–gels. The sol–gel network can be formed in the presence of the biomolecules (*in situ* entrapment) or it is loaded thereafter (8, 9). The latter can be achieved by adsorption or covalent binding of the biomolecules to the sol–gel support and was also

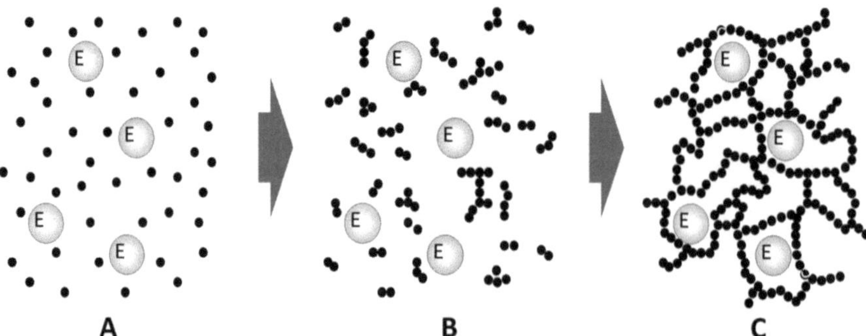

Fig. 18.1. Entrapment of biomolecules in sol–gels: due to the hydrolysis of the silica alkoxides and their subsequent condensation, colloidal particles are formed. Proceeding condensation causes the formation of a three-dimensional network (gelation). If biomolecules like enzymes are present during the gelation, they become tightly encaged in the formed network.

successfully demonstrated for multi-enzyme systems (10). But adsorption often encounters insufficient process stability due to the leaching of the biocatalyst, which in turn can be prevented by covalent binding of the biomolecules to the sol–gel network. However, covalent binding often entails some implication on the biocatalyst stability, hence it is not applicable to all enzymes. A remedy for the above-mentioned drawbacks is the *in situ* entrapment when the bioactive substances are added to the precursors and physically entrapped inside the pores in the course of the sol–gel formation (**Fig. 18.1**).

The advantage of the *in situ* entrapment is that the bioactive substances are tightly encaged to an extent that normally cannot be achieved by simple adsorption or covalent binding (8, 11). The cage prevents the leaching of the biomolecules and offers protection against detrimental thermal effects (12). At the same time the beneficial high surface area to volume ratio and the nanoporous structure account for a sufficiently high mass transport. The sol–gel matrices can be considered as chemically inert, and if not modified as hydrophilic (due to residual silanol groups), biocompatible and inexpensive to produce. They also do not show swelling in solvents compared to many organic polymers. Typically enzymes entrapped in sol—gels show a high stability which can be best explained by their spatial confinement: unfolding and denaturation are less likely to occur inside a confined space due to missing space needed for the unfolding of the protein chain – a simple but highly effective principle (13, 14). Based on the reaction mechanism in the sol–gel process, no reaction between enzyme and silica matrix will occur, therefore negligible loss of enzyme activities during the encapsulation step can be expected. But admittedly since the sol–gel formation comes along with the release of alcohol, at least some enzymes are adversely affected. Nonetheless, the sol–gel encapsulation of broad array of bioactive substances (like enzymes, antibodies, nucleic acids, and living cells) emerged out as versatile method to yield high-performance biocatalysts for applications in the biocatalytical synthesis of specialty or commodity chemicals, optical and electrochemical biosensing, bio-affinity chromatography, and controlled released systems (8).

Extensive research efforts resulted in a plentitude of different approaches to fine-tune the sol–gel matrix. The combination of inorganic and organic moieties may result in composites with unique properties, especially regarding the pore size distribution, optical transmittance (if used for an optical biosensor), high resistance against brittleness, conductivity, and the performance of the entrapped biomolecule. Such organically modified sol–gels are frequently referred as ormosils (8). Based on sol–gel technology, the easiest way to introduce organic components like alkyl chains or other functional groups is by using low molecular weight siloxanes carrying the desired functionality (e.g., alkyl chains or functional groups like –COOH, NH_2, –SH, and –OH) or by the incorporation of appropriately functionalized polymers (15, 16). Another possibility is the admixing of metallic nanoparticles or conductive polymers, e.g., if a conductive entrapment matrix for an enzymatic redox system is desired (17–19).

This chapter introduces the method of immobilizing a lipase in an alkyl-modified sol–gel for the synthesis of biodiesel in order to demonstrate the procedure of the method and at the same time the great potentials of the sol–gel immobilization technology.

2. Materials

1. Solution I (buffer): 0.1 M Potassium phosphate buffer, pH 7.0.
2. Solution II (enzyme solution): 1.25 mg/mL Enzyme protein in potassium phosphate buffer. For the results given below, a highly active lipase preparation from *Thermomyces lanuginosa* (Chirazyme L8-lyo, protein content of 35 wt% and a lipolytic activity of > 3,000 kU/g) was used, which was kindly provided by Roche Diagnostic GmbH.
3. Solution III (protein standard): 100 μg/mL Protein standard (bovine serum albumin (BSA)) in deionized water. BSA and Bradford reagent are purchased at Sigma-Aldrich Co (St. Louis, MO).
4. Solution IV (substrate solution for the determination of the lipolytic activity): Emulsion of 50 mM tributyrin in 1 mM Tris–HCl buffer of pH 7.5 with 0.1 mM $CaCl_2$ and 0.1 mM NaCl.
5. Solution V (titrant for determination of lipolytic activity): 50 mM NaOH.
6. Solution VI (sol–gel catalyst): 1 M NaF in water.
7. Solution VII (emulsifier): 4% (w/v) Polyvinyl alcohol (PVA, 15.000 Da) in water.
8. Solution VIII (substrate solution for transesterification): 45 g/L Canola oil, 5 g/L methanol, and 7.5 g/L *n*-decane in *n*-hexane.

3. Methods

3.1. Characterization of the Lipase Preparation

Lipases are efficient catalysts for esterifications, transesterifications, and hydrolysis reactions and have consequently enormous importance in the bioindustry (20). Many lipases are commercially available, with each being developed or proven for specific applications, and so the single preparation will vary tremendously with regard to purity, protein content, specific activity as well as the way the formulation is provided (as liquid or powder) containing various stabilizing reagents (**Note 1**). Normally, an in-depth characterization of the lipase preparation (solution II) is not needed, but at least the protein content and the lipolytic activity should be determined. The activity of the free enzyme in the aimed reaction is also of interest and can be determined as described later (*see* **Section 3.4**).

3.1.1. Bradford Assay for Determination of Protein Concentration

The Bradford assay utilizes a colorimetric reagent for the detection and quantification of total protein (21). Protein in solution binds to the reagent (Coomassie Brilliant Blue G250) accompanied by a spectral shift from 465 to 595 nm. The spectral shift is proportional to the amount of protein present in the solution. Sample protein concentration is determined by calibration with BSA as standard protein. The applied calibration range is between 0 and 20 μg/mL (solution III as sample solution). The sample solution must be diluted appropriately to a final volume of 500 μL, transferred into a 1.5-mL cuvette followed by the addition of 500 μL Bradford reagent (ready-to-use preparation obtainable from chemical supplier). The absorption at 595 nm is measured with a UV–Vis spectrophotometer. The protein content of the enzyme preparation will be determined by the solubilization of a predefined amount of the enzyme powder in water or buffer (e.g., 10 mg in 1 mL). If necessary, insoluble debris has to be removed by centrifugation prior to the dilution of the sample (**Note 2**).

3.1.2. Tributyrin Assay for the Determination of the Lipolytic Activity

The lipase-catalyzed hydrolysis of tributyrin results in the release of butyric acid and is therefore accompanied by a pH drop. The amount of butyric acid can be determined via titration, most suitable with an autotitrator:

1. The reaction chamber of the autotitrator is filled with 10 mL of the substrate solution (solution IV, emulsion of tributyrin in Tris–HCl buffer) and vigorously stirred. The reaction will not start unless the pH has stabilized (pH 7.5).
2. A suitable amount of the lipase solution (solution II, e.g., 10 μL) is added and the titration with 50 mM NaOH (solution V) is started in order to keep the pH stable. Reaction should last for 30 min and the titrant should be added at a constant rate. A declining rate indicates that the enzyme amount is too high and with proceeding of the reaction, enzyme molecules are no more saturated with substrate.
3. The amount of released butyric acid is calculated from the volume of 50 mM NaOH added at a constant rate and correlated to the protein amount present in the assay. The specific

activity of 1 U/g denotes the release of 1 μmol butyric acid per minute catalyzed by 1 g of protein.

3.2. Entrapment of a Lipase in Alkyl-Modified Sol–Gel Silica

Lipase entrapment in alkyl-modified sol–gel has proven to make highly active biocatalysts available for non-aqueous biocatalysis (3, 22). Previous works showed a very low residual activity after sol–gel entrapment of lipases without alkyl modification (23). This was surprising since lipases typically must be considered as robust and should not be prone to be inactivated during the sol–gel process. So it was concluded that the reason for this is given by the intrinsic nature of the lipase: It is well known that most lipases become activated by hydrophobic/hydrophilic interfaces, which induce a structural change (uncovering the active site) to render the lipase molecule active (20). In the inactive state, a "lid" is covering the active site, which is therefore inaccessible to substrates. Unmodified silica gels are hydrophilic due to some remaining silanol groups and not able to trigger the needed structural change. It was proven that the implementation of hydrophobic alkyl chains to lipase containing sol–gels results in a significant higher activity, even exceeding that of the free dispersed enzyme (3, 22, 24). Obviously the interaction of a lipase with alkyl-modified sol–gels stimulates the enzyme in a similar manner as a hydrophobic/hydrophilic interface does. The sol–gel technology allows adjusting the balance between the hydrophilic and the hydrophobic nature of the sol–gel materials to provide a favorable microenvironment for the enzyme and the substrate/product in the interest of a high catalytic activity. Some researchers (3) reported a several hundred-fold enhancement of lipase activity compared to the free enzyme in the esterification of octanol and lauric acid in isooctane. At the same time, the lipase benefits from the spatial confinement in the sol–gel, cognizable by an enhanced thermal stability (5).

3.2.1. Sol–Gel Process

The sol–gel process involves a transition of a system from a sol, a relatively low viscous liquid, into a solid gel phase by the preceding condensation and cross-linking of the colloidal particles detectable by a sharp increase of viscosity (8). As precursors for a sol–gel reaction, typically metal alkoxides like tetramethoxysilane (TMOS) or tetraethoxysilane (TEOS) are used. First, the alkoxide is easily hydrolyzed by water if catalyzed by an acid or a base. The resulting $Si(OH)_4$ tetrahedra are subjected to further condensation yielding siloxane bonds. In the following polycondensation, additional silanols get linked. As soon as enough Si–O–Si bonds are present, they will form the sol comprising colloidal particles. The size of these colloidal particles and the degree of cross-linking depend on the concentration of the alkoxide as well as on the pH. The further cross-linking of the particles, gelation, results in the formation of a three-dimensional network in which

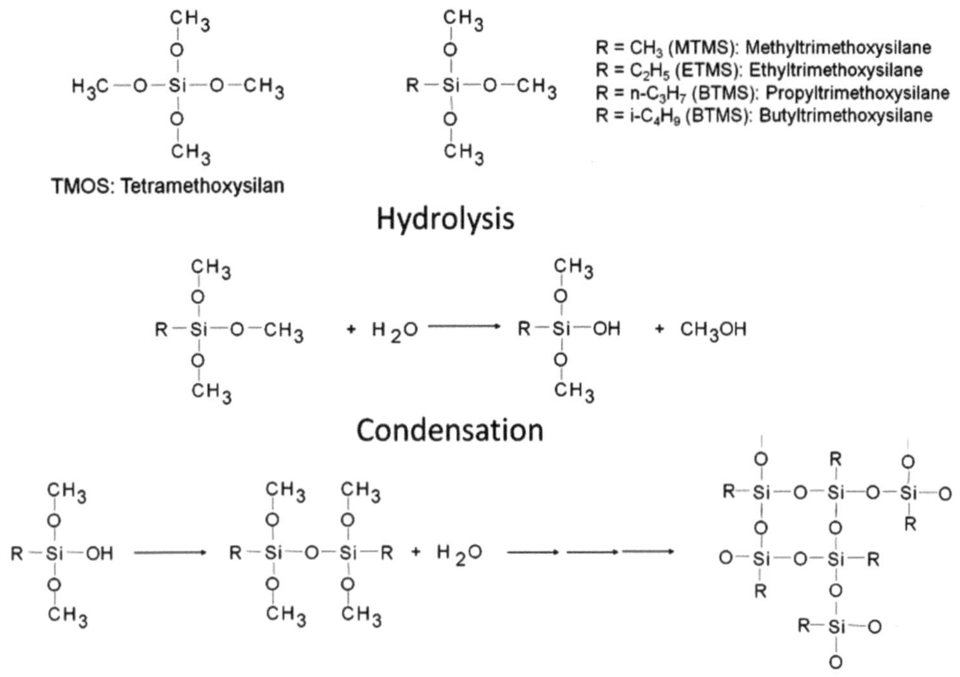

R = CH_3 (MTMS): Methyltrimethoxysilane
R = C_2H_5 (ETMS): Ethyltrimethoxysilane
R = n-C_3H_7 (BTMS): Propyltrimethoxysilane
R = i-C_4H_9 (BTMS): Butyltrimethoxysilane

TMOS: Tetramethoxysilan

Hydrolysis

Condensation

Fig. 18.2. Reaction scheme of the sol–gel process with (alkyl-modified) silica alkoxides.

the biomolecules are entrapped. The whole process is depicted in **Figs. 18.1** and **18.2**. By further drying, the "gel" is converted into dense silica glass, which is often referred to as xerogel (11). The characteristics of the sol–gel are pre-determined by the nature of the precursors used, the process conditions, and the use of additives. If desired, the sol can be cast into a mold to obtain solid objects of a precisely defined shape as they may be used in optical biosensors (7, 11).

3.2.2. Sol–Gel Preparation

1. First, 2 mL of the lipase containing solution (solution II) is transferred into a 10-mL screwable vial. Subsequently 500 μL of an aqueous solution of sodium fluoride (solution VI), 1 mL solution of polyvinyl alcohol (solution VII) in water, and varying amounts of sol–gel precursors (tempered to 4°C) are added as specified in **Table 18.1**. In this way, silica glasses of differently balanced hydrophobicity and hydrophilicity are generated (**Note 3**).
2. The mixtures are vigorously shaken until the gelation starts. Then the mixture is placed on ice for approximately 5 min and finally the mixture is left at room temperature for 24 h.
3. The obtained sol–gels are transferred into a Petri dish, dried at 35°C for 24 h, and pestled (**Note 4**). After the determination of weight, the sol—gels are suspended in 10 mL water for 2 h to remove loosely adsorbed protein. The protein

Table 18.1
Pipetting scheme for the preparation of lipase containing sol–gels modified by the introduction of alkyl chains

Sol–gel	Enzyme solution (mL)	NaF (mL)	PVA (mL)	Deionized H_2O (mL)	Sol–gel precursors
TMOS	2	0.5	1	0.82	4.425 mL TMOS
MTMS	2	0.5	1	0.82	4.285 mL MTMS
ETMS/TMOS	2	0.5	1	0.82	3.96 mL ETMS + 0.74 mL TMOS
PTMS/TMOS	2	0.5	1	0.82	4.375 mL PTMS + 0.74 mL TMOS
BTMS/TMOS	2	0.5	1	0.82	4.78 mL BTMS + 0.74 mL TMOS

content in the washing water is quantified to calculate the degree of immobilization (*see* **Sections 3.1** and **3.3**).

4. The wet sol–gels are washed again over a glass frit with 100 mL water, dried at 40°C for 12 h, and stored at room temperature. In case of longer alkyl chains, it is recommended to wash the sol–gel with water-miscible organic solvents to facilitate the removal of methanol.
5. The yield of sol–gel is determined as the ratio of total amount of used material to total mass of obtained silica glass (considering the amount of water and the methanol released during the hydrolysis of the siloxanes). Typically yields of more than 80% are achieved (**Note 5**).

3.2.3. Sol–Gel Characterization

Once formed the nanoporous materials can be characterized by a variety of methods. If the sol–gel is not used in a defined shape, it will be grinded to the desired particle size. Size of the sol–gel particles can be determined by particle sizing systems based on light scattering (4). Typically sol–gel particles show a broad size distribution with an average diameter of 30 μm (4). Besides that, the most interesting parameter to know is the pore size, volume and distribution, and consequently the accessible surface area (surface-to-weight ratio). Predominantly used characterization techniques are gas adsorption, X-ray diffraction, and electron microscopy (4, 25). Electron microscopy is recommended for an inspection of the topography of the nanoporous material. One predominant method is the gas adsorption measurement following the *Brunauer–Emmett–Teller* (BET) theory to determine the surface area and the *Barret–Joyner–Halenda* (BJH) theory for the size distribution (26): A porous solid is capable of adsorbing a large volume of condensable gas after the gas molecules has penetrated the porous material, resulting in a gradual decline of the gas pressure (which equals to the amount of adsorbed gas) if the porous material is enclosed in a closed chamber at a certain temperature. Analysis of the pore structure becomes possible by the evaluation of isotherms (which refers to the amount of adsorbed gas depending on the temperature). For the entrapment of a lipase in a tetramethoxysilane/*i*-butyltrimethoxysilane (TMOS/*i*BTMS), sol–gel at a specific surface area of around 200 m^2/g was reported, but even values up to 860 m^2/g have been achieved (4). Similar values were obtained by Kuncova et al. (22), who also used the sol–gels made from tetramethoxysilane, methyl-trimethoxysilane, propyltrimethoxysilane, and (3-aminopropyl)triethoxysilane. The pore diameters of the prepared sol–gels were calculated by the BJH method and typically range for this type of sol–gels between 2 and 10 nm (11, 22, 27).

3.3. Immobilization Efficiency

The efficiency of the immobilized biocatalysts can be appraised only if compared to the free non-immobilized enzyme. This requires the quantitative determination of the specific catalytic activity, which has to be correlated with the protein concentration (*see* **Sections 3.4** and **3.5**). Thereby it has to be differentiated between the specific activity (product formation per milligram protein) of the native enzyme, the specific activity of the immobilized protein (product formation per milligram protein), and the apparent activity of the immobilizate (product formation per gram sol–gel). The evaluation of those activities makes it necessary to know the amount of enzyme protein entrapped within the sol–gel. Since the enzyme solution is directly mixed with the sol–gel precursors, the amount of immobilized protein has to be determined by washing the dried sol–gel with water to remove loosely adsorbed protein. The protein concentration in the washing water is determined with the Bradford assay (*see* **Section 3.1**) and balanced with the total amount of applied protein to give the degree of immobilization (fraction of total protein getting immobilized). Typically in the sol–gel process, immobilization degrees of 70–90% can be achieved (**Table 18.2**). Similar values are reported in literature (3, 23).

3.4. Transesterification Catalyzed by Sol–Gel-Immobilized Lipase

In industrial biotransformation, the reactions are frequently carried out in organic solvents since they allow a better solubility/stability of substrate and/or product molecules as well as a beneficial shift of the thermodynamic equilibrium (28). Lipases, for example, are able to catalyze the reverse reaction if used in an organic solvent. They can be used in monophasic, either aqueous or organic, media or in a combination of both, i.e., biphasic systems (20). Since transesterification and esterification reactions are of particular interest for the synthesis of fine chemicals, much

Table 18.2
Specific and apparent activities, protein content, and immobilization degree obtained after immobilization of the lipase from *T. lanuginosa* in sol–gels made of TMOS, MTMS, ETMS, PTMS, or *i*-BTMS

Sol–gel type	Specific activity (g/L h $mg_{Protein}$)	(g/L h g_{Immo})	Immobilization degree (%)	Protein content in sol–gel (μg/mg)
Free lipase	1.46 ± 0.09	n.a.	n.a.	n.a.
TMOS	0.78 ± 0.10	0.54 ± 0.07	92.3 ± 3.9	0.69 ± 0.03
MTMS	2.24 ± 0.33	2.23 ± 0.33	86.8 ± 4.1	1.00 ± 0.04
ETMS/TMOS	83.09 ± 13.51	80.68 ± 13.12	79.3 ± 11.7	0.97 ± 0.11
PTMS/TMOS	146.44 ± 37.46	86.40 ± 22.09	69.2 ± 8.6	0.59 ± 0.05
BTMS/TMOS	143.38 ± 24.86	143.53 ± 24.89	78.9 ± 10.9	1.00 ± 0.11

research efforts were dedicated to this field. For the characterization of immobilized enzymes, especially if different methods shall be compared, it has worked out to deploy a relevant model reaction. Lipases are frequently used in the food processing or personal care industry for the transesterification of triglycerides (20). They can also be used for the synthesis of biodiesel by transesterification of vegetable oil with methanol (29), the chosen reaction model in this chapter. Such reactions are typically carried out solvent free, but here for the easiness of the reaction, the substrates are solubilized in *n*-hexane to lower the viscosity.

3.4.1. Activity Assay

1. An appropriate amount of the dry sol–gel (typically 350 mg) or the free powder-like lipase (20 mg) is weighted into 10-mL screwable glass vials (**Notes 6**, **7**, and **8**).
2. The reaction is started by the addition of 5 mL substrate solution (solution VII, canola oil, and methanol) and the vials are vigorously shaken at 35°C temperature. Typically in a reaction time of 20 h, a conversion of around 90% can be achieved.
3. Samples of 40 μL are withdrawn in suitable time intervals, 1:5 diluted, and transferred into GC vials. Thereby it must be assured that the suspended particles are fully sedimented before the sample is taken. Formation of fatty acid methyl esters is quantified and the initial reaction rate is calculated and correlated with the used catalyst amount to give the specific activity.

3.4.2. GC-FID Analysis

Vegetable oils and especially the reaction mixture obtained in the course of the transesterification contain a multitude of different compounds. Various fatty acids are esterified with glycerol in mono-, di-, and triglycerides in all possible combinations, while the fatty acids are present free or as methyl esters. The quantitative analysis of each single molecule would encounter a nonjustifiable effort, and hence it has become a common practice to analyze the different molecules in those transesterification mixtures after being classified in certain groups (30, 31). Fatty acids are classified according to their chain length, e.g., as C16 and C18 fatty acids and likewise the corresponding esters as C16 and C18 fatty acid methyl esters. Mono-, di-, and triglycerides are classified regardless of the position and nature of the esterified fatty acids, though the length of the fatty acid should be considered.

Prerequisite for this analytical method are operational conditions that assure the concomitant elution of all molecules belonging to a certain group at the same time. This can be achieved by the deployment of a 15-m fused silica capillary column with a stationary phase made of 5% phenyl/95% methylpolysiloxane (as marketed under names like DB-5, HP-5, VF-5, and BPX5) with an inner diameter (ID) of 0.25 mm and a film thickness (df) of 0.25 μm. The following operational conditions and procedures are recommended:

1. Injection and detection temperature, 365°C; carrier gas and flow rate: helium and 8 mL/min; initial temperature 70°C, hold for 1 min and increase at the rate of 45°C/min up to 360°C; total length 15 min.
2. For the verification of the undertaken classification, pure standards of selected fatty acids, fatty acid methyl esters, mono-, di-, and triglycerides should be injected.
3. Calibration can be conducted with canola oil in *n*-hexane in a concentration range from 0.1 to 10 g/L with 1.5 g/L *n*-decane as internal standard. For the quantification of the formed fatty acid methyl esters, the canola oil was transesterified with sodium methanolate as described elsewhere (32). The obtained methyl esters were used for calibration in *n*-hexane with concentrations ranging from 0.1 to 10 g/L using 1.5 g/L *n*-decane as internal standard.

3.5. Specific Activities of the Sol–Gel-Immobilized Lipase Compared to the Free Enzyme

The specific and apparent activities of the lipase from *T. lanuginosa* obtained after entrapment of varying enzyme amounts in differently modified sol–gels are summarized in **Table 18.2**. Interestingly the immobilization of the lipase in unmodified sol–gel resulted in an even lower specific activity compared to the free enzyme. Sol–gels made of MTMS (methyltrimethoxysilane) allowed a moderate enhancement of the activity. With increasing chain length over MTMS and ETMS (ethyltrimethoxysilane), a drastically increasing specific activity was observed (**Fig. 18.3**). A considerable activation – a 100-fold

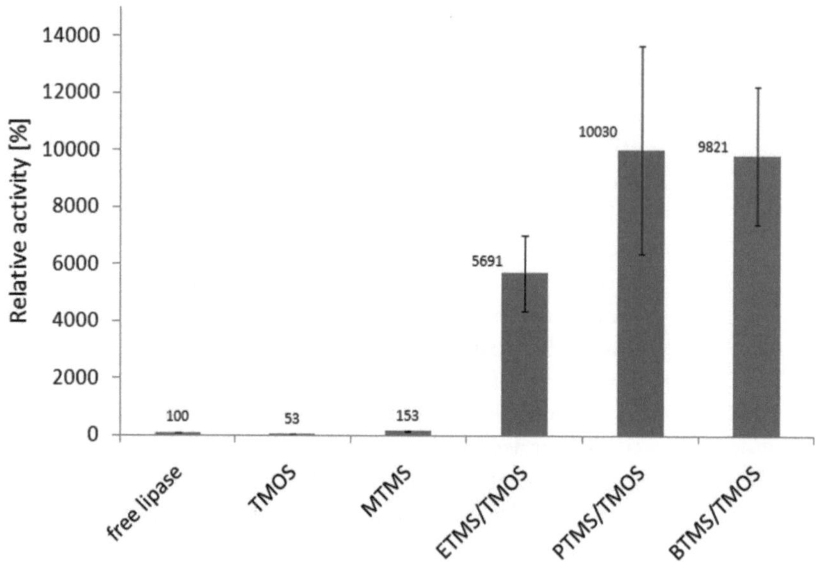

Fig. 18.3. Enhancement of lipase activity by entrapment in alkyl-modified sol–gels compared to the specific activity of the non-immobilized enzyme.

enhancement – was observed in those sol–gels composed of TMOS and PTMS (propyltrimethoxysilane) or of TMOS and *i*BTMS (*i*-butyltrimethoxysilane). Even though the specific activity related to the amount of protein is the same for both TMOS/PTMS and TMOS/*i*BTMS, the apparent activity differs due to the different yield of immobilization. Nonetheless the results clearly indicate that the lipase obviously benefits from a hydrophobic microenvironment. The enzyme performance is remarkable improved by a simple modification of the sol–gel through the introduction of alkyl chains. Admittedly these data do not represent the outcome of an extensive optimization. Numerous highly sophisticated procedures for the entrapment of lipase in sol–gels have been published which cannot be reviewed here. The main objective of this chapter is to demonstrate the potential of the easy tailorable sol–gel entrapment, which is applicable to all kinds of enzymes.

4. Notes

1. Enzyme protein concentration to be used in sol–gel entrapment strongly depends on the purity and activity of the formulation used, which has to be determined experimentally.
2. Occasionally it turns out to be difficult to solubilize an enzyme powder completely. If so, it may help to lower the

concentration or to change the pH and buffer. Remaining insoluble particles should be removed by centrifugation.

3. Since several steps must be carried out following this protocol, it must be taken into account that every single step may contribute to the final standard deviation of the specific activity. For this reason, experiments should be carried out at least in triplicate.
4. If the sol–gels are not thoroughly dried or washed, some residual alcohol might be present in the sol–gel matrix, leading to a lower enzyme activity and/or undesired side products.
5. The yield of sol–gels can be adversely affected if some material sticks to the wall of the reaction tubes used during the gelation. This becomes particularly a problem if alkoxides with longer alkyl chains are used.
6. Freshly prepared sol–gels show a certain affinity to apolar substrates and products, hence an initial decrease of substrate concentration with the absence of corresponding amounts of products can be explained. Normally this lag phase lasts less than 30 min after the reaction has started.
7. Sol–gels can be reused, but it has to be considered that apolar substrates and products might be adsorbed to the sol–gel material. For this reason, it is necessary to wash the sol–gel carefully with an organic solvent until no substrate or product is detected in the washing solvent prior to be re-used.
8. Sol–gels made by this protocol tend to be brittle and the use of a stirring bar might "grind" the particles. This may turn out to be a problem for certain reactor designs and may hamper the withdrawal of samples. Therefore it is recommended to avoid excessive shear forces during the reaction.
9. The protocol given above can be easily applied to other enzymes, especially lipases or esterases. If low activities are obtained, it might be possible that the enzyme is adversely affected during the sol–gel formation, e.g., by extreme pH values and high levels of alcohol in the enzyme's microenvironment. Moreover, the silicate matrix exposes a negative net charge, hence unfavorable interactions with enzyme molecules or charged substrates/products may occur.

References

1. Ansorge-Schumacher, M. B. (2008) Immobilization of biological catalysts. *Handbook of Heterogeneous Catalysis*, 2nd Edn., Vol. 1, pp. 644–655.
2. Hanefeld, U., Gardossi, L., and Magner, E. (2009) Understanding enzyme immobilization. *Chem. Soc. Rev.* **38**, 453–468.

3. Reetz, M. T., Tielmann, P., Wiesenhoefer, W., Koenen, W., and Zonta, A. (2003) Second generation sol-gel encapsulated lipases: Robust heterogeneous biocatalysts. *Adv. Synth. Catal.* **345**, 717–728.
4. Noureddini, H., and Gao, X. (2007) Characterization of sol-gel immobilized lipases. *J. Sol-Gel Sci. Technol.* **41**, 31–41.
5. Dunker, A. K., and Fernandez, A. (2007) Engineering productive enzyme confinement. *Trends Biotechnol.* **25**, 189–190.
6. Keeling-Tucker, T., Rakic, M., Spong, C., and Brennan, J. D. (2000) Controlling the material properties and biological activity of lipase within sol-gel derived bioglasses via organosilane and polymer doping. *Chem. Mater.* **12**, 3695–3704.
7. Gupta, R., and Chaudhury, N. K. (2007) Entrapment of biomolecules in sol-gel matrix for applications in biosensors: Problems and future prospects. *Biosens. Bioelectron.* **22**, 2387–2399.
8. Kandimalla, V. B., Tripathi, V. S., and Ju, H. (2006) Immobilization of biomolecules in sol-gels: Biological and analytical applications. *Crit. Rev. Anal. Chem.* **36**, 73–106.
9. Lan, E. H., Dunn, B., and Zink, J. I. (2005) Nanostructured systems for biological materials. *Methods Mol. Biol.* **300**, 53–79.
10. El-Zahab, B., Jia, H., and Wang, P. (2004) Enabling multienzyme biocatalysis using nanoporous materials. *Biotechnol. Bioeng.* **87**, 178–183.
11. Dave, B. C., Dunn, B., Valentine, J. S., and Zink, J. I. (1994) Sol-gel encapsulation methods for biosensors. *Anal. Chem.* **66**, 1120A–1127A.
12. Noureddini, H., Gao, X., Joshi, S., and Wagner, P. R. (2002) Immobilization of Pseudomonas cepacia lipase by sol-gel entrapment and its application in the hydrolysis of soybean oil. *J. Am. Oil Chem. Soc.* 79, 33–40.
13. Wang, P. (2006) Nanoscale biocatalyst systems. *Curr. Opin. Biotechnol.* **17**, 574–579.
14. Wang, P., Dai, S., Waezsada, S. D., Tsao, A. Y., and Davison, B. H. (2001) Enzyme stabilization by covalent binding in nanoporous sol-gel glass for nonaqueous biocatalysis. *Biotechnol. Bioeng.* 74, 249–255.
15. Schmidt, H., Scholze, H., and Kaiser, A. (1984) Principles of hydrolysis and condensation reaction of alkoxysilanes. *J. Non-Cryst. Solids* **63**, 1–11.
16. Wilkes, G. L., Orler, B., and Huang, H. H. (1985) "Ceramers": Hybrid materials incorporating polymeric/oligomeric species into inorganic glasses utilizing a sol-gel approach. *Polym. Prepr. (Am. Chem. Soc., Div. Polym. Chem.)* **26**, 300–302.
17. Park, T., Iwuoha, E. I., Smyth, M. R., and MacCraith, B. D. (1996) Sol-gel-based amperometric glucose biosensor incorporating an osmium redox polymer as mediator. *Anal. Commun.* **33**, 271–273.
18. Siu, E., Won, K., and Park, C. B. (2007) Electrochemical regeneration of NADH using conductive vanadia-silica xerogels. *Biotechnol. Prog.* **23**, 293–296.
19. Li, W., Yuan, R., Chai, Y., Zhou, L., Chen, S., and Li, N. (2008) Immobilization of horseradish peroxidase on chitosan/silica sol-gel hybrid membranes for the preparation of hydrogen peroxide biosensor. *J. Biochem. Biophys. Methods* **70**, 830–837.
20. Saxena, R. K., Ghosh, P. K., Gupta, R., Davidson, W. S., Bradoo, S., and Gulati, R. (1999) Microbial lipases: Potential biocatalysts for the future industry. *Curr. Sci.* 77, 101–115.
21. Bradford, M. M. (1976) A rapid and sensitive method for the quantitation of microgram quantities of protein utilizing the principle of protein-dye binding. *Anal. Biochem.* **72**, 248–254.
22. Kuncova, G., Szilva, J., Hetflejs, J., and Sabata, S. (2003) Catalysis in organic solvents with lipase immobilized by sol-gel technique. *J. Sol-Gel Sci. Technol.* **26**, 1183–1187.
23. Reetz, M. T., Zonta, A., and Simpelkamp, J. (1996) Efficient immobilization of lipases by entrapment in hydrophobic sol-gel materials. *Biotechnol. Bioeng.* **49**, 527–534.
24. Chen, J., and Lin, W. (2003) Sol-gel powders and supported sol-gel polymers for immobilization of lipase in ester synthesis. *Enzyme. Microb. Technol.* **32**, 801–811.
25. Lei, C., Soares, T. A., Shin, Y., Liu, J., and Ackerman, E. J. (2008) Enzyme specific activity in functionalized nanoporous supports. *Nanotechnology* **19**, 125102/1–125102/9.
26. Sing, K. S. W. (1998) Adsorption methods for the characterization of porous materials. *Adv. Colloid Interface Sci.* **76–77**, 3–11.
27. Wei, Y., and Qiu, K. (2000) Synthesis and biotechnological applications of vinyl polymer-inorganic hybrid and mesoporous materials. *Chin. J. Polym. Sci.* **18**, 1–7.
28. Zaks, A., and Klibanov, A. M. (1984) Enzymatic catalysis in organic media at 100 degrees C. *Science* **224**, 1249–1251.
29. Parawira, W. (2009) Biotechnological production of biodiesel fuel using biocatalysed transesterification: A review. *Crit. Rev. Biotechnol.* **29**, 82–93.
30. Knothe, G. (2001) Analytical methods used in the production and fuel quality assessment of biodiesel. *Trans. ASAE* **44**, 193–200.

31. Mittelbach, M., Roth, G., and Bergmann, A. (1996) Simultaneous gas chromatographic determination of methanol and free glycerol in biodiesel. *Chromatographia* **42**, 431–434.

32. Martyanov, I. N., and Sayari, A. (2008) Comparative study of triglyceride transesterification in the presence of catalytic amounts of sodium, magnesium, and calcium methoxides. *Appl. Catal. A* **339**, 45–52.

Index

P. Wang (ed.), *Nanoscale Biocatalysis*, Methods in Molecular Biology 743,
DOI 10.1007/978-1-61779-132-1, © Springer Science+Business Media, LLC 2011

MIX
Papier aus verantwortungsvollen Quellen
Paper from responsible sources
FSC® C105338

If you have any concerns about our products,
you can contact us on
ProductSafety@springernature.com

In case Publisher is established outside the EU,
the EU authorized representative is:
Springer Nature Customer Service Center GmbH
Europaplatz 3, 69115 Heidelberg, Germany

Printed by Libri Plureos GmbH
in Hamburg, Germany